THE
PROFESSIONAL ENGINEERING
CAREER DEVELOPMENT SERIES

CONSULTING EDITORS

Dr. John J. McKetta, Jr., Executive Vice Chancellor for Academic Affairs, The University of Texas System

Dr. Maurits Dekker

EDITORIAL ADVISORY COMMITTEE

Dr. Walter O. Carlson, Acting Dean of Engineering College, Georgia Institute of Technology

Mr. R. E. Carroll, Director, Continuing Engineering Education, The University of Michigan

Mr. William W. Ellis, Director, Post College Professional Education, Carnegie-Mellon University

Dr. Gerald L. Esterson, Director, Division of Continuing Professional Education, Washington University

Dr. L. Dale Harris, Associated Dean, College of Engineering, The University of Utah

Dr. James E. Holte, Director, Continuing Education in Engineering and Science, University of Minnesota

Dr. Russell R. O'Neill, Associate Dean and Professor, School of Engineering and Applied Science, University of California at Los Angeles

ENGINEERING PROFESSION
ADVISORY GROUP

Professional Engineering Career Development Series

WATER QUALITY ENGINEERING FOR PRACTICING ENGINEERS

W. Wesley Eckenfelder, Jr.

Department of Environment and
Water Resources Engineering
Vanderbilt University

BARNES & NOBLE, INC. NEW YORK
Publishers · Booksellers · Since 1873

Distributed

In Canada
by the Ryerson Press, Toronto

In Australia and New Zealand
by Hicks, Smith & Sons Pty. Ltd.,
Sydney and Wellington

In the United Kingdom, Europe,
and South Africa
by Chapman & Hall Ltd., London

PREFACE

It has only been in the past decade that water-quality engineering has begun to emerge from an art to a science. The ever increasing need for water-pollution control has stimulated research in new and improved methods of wastewater treatment and in developing a better understanding of the short and long-term effects of wastewaters on the aquatic environment. In order to fill the expanding needs for trained personnel in water quality, increasing numbers of scientists and engineers from many disciplines are being drawn into the field.

It is the purpose of this book to present a concise summary of the present principles and theories on water-pollution control, to indicate processes and treatments applicable to specific sewage and industrial wastewater problems, to define the significant parameters in water-quality engineering, and to develop design procedures for the wastewater-treatment processes in most common use today.

This book should be useful as an introductory text for engineers from other disciplines engaged in the water-quality field as well as providing engineering guidelines for the solution of particular problems.

For in-depth study of specific areas, the reader is referred to the cited books and references.

Much of the material presented in this book was drawn from the author's notes and course manuals prepared for the following courses: (1) Water-Pollution Control in the Chemical Industry, sponsored by the Manufacturing Chemists Association, and (2) Advanced Water-Pollution Control, offered by the College of Engineering and the Engineering Foundation of The University of Texas at Austin and by the Summer Institute in Water Pollution Control at Manhattan College.

Some material was also drawn from a series of manuals on water-pollution control prepared for the World Health Organization.

Acknowledgments

The author acknowledges the reproduction of many figures and tables taken from his reports and technical papers.

The encouragement and assistance of the faculty and staff of the Environmental Health Engineering group at The University of Texas at Austin is gratefully acknowledged. A word of appreciation is due to the several colleagues and students who checked the manuscript for technical accuracy and to Kathy Pickens for her expert typing of the manuscript.

CONTENTS

WATER QUALITY ENGINEERING FOR PRACTICING ENGINEERS

1

General Concepts of Water-Quality Management

Over the past decade, water-pollution control has progressed from an art to a science. Increased emphasis has been placed on the removal of secondary pollutants such as nutrients and refractory organics and on water reuse for industrial and agricultural purposes. This, in turn, has generated both fundamental and applied research, which has improved both the design and operation of wastewater-treatment facilities.

Solving water-pollution problems today involves a multidisciplinary approach in which the required water quality is related to agricultural, municipal, recreational, and industrial requirements. In many cases a cost-benefit ratio must be established between the benefit derived from a specified water quality and the cost of achieving that quality.

Wastewaters emanate from four primary sources: (1) municipal sewage, (2) industrial wastewaters, (3) agricultural runoff, and (4) storm-water and urban runoff.

Estimating municipal wastewater flows and loadings can be done in one of several ways, based on knowledge of past and future growth plans for the community, sociological patterns, and land-use planning.

1. Population-prediction techniques. Several mathematical techniques are available for estimating population growth. Caution should be employed in the use of these procedures,

particularly in areas subject to rapid industrial expansion, rapid suburban development, and changing land-use patterns.

2. Saturation population from zoning practice. Percentages of a saturation population can be estimated for fully developed areas based on zoning restrictions (single dwelling residential, multiple dwelling residential, commercial, etc.).

Provision should be included for infiltration in the case of separate sewers, and storm flows in the case of combined sewers.

The average characteristics of domestic sewage are listed in Table 4.1. The estimated sewage contributions from commercial and industrial establishments are summarized in Table 4.4.

As municipal and industrial wastewaters receive treatment, increasing emphasis is being placed on the pollutional effects of urban and agricultural runoff. The range of concentration of pertinent characteristics in these wastewaters is given in Table 1.1. Present research on storm-water treatment consid-

Table 1.1. Pollution from Urban and Agricultural Runoff

Constituent	Urban Runoff[a] (Storm Water)	Agricultural Runoff[b]
Suspended solids, mg/l	5–1200	—
Chemical oxygen demand (COD), mg/l	20–610	—
Biological oxygen demand (BOD), mg/l	1–173	—
Total phosphorus, mg/l	0.02–7.3	0.1–0.65
Nitrate nitrogen, mg/l	—	0.03–5.0
Total nitrogen, mg/l	0.3–7.5	0.5–6.5
Chlorides, mg/l	3–35	—

[a] From Weibel et al. [4].
[b] From Sylvester [5].

ers large holding basins in which the storm waters are treated in the municipal facility after the storm—an in situ treatment by screening, sedimentation, chlorination, and so on. In the future, water-quality management in highly urbanized areas will have to consider storm water as a major pollutant.

Agricultural runoff is a major contributor to eutrophication in lakes and other natural bodies of water. Effective control measures have yet to be developed for this problem. Runoff of pesticides is also receiving increasing attention.

A procedure for the development of an effective water-quality management program is shown in Fig. 1.1. Such a program is directed toward establishing the most economic long-range solution for specified water-quality needs.

After establishing the water-quality criteria, design considerations should include combined or separate treatment, ocean outfalls in coastal areas, and flexibility for expansion of the facilities to upgrade the effluent quality as the loading to the facilities is increased.

WATER-QUALITY STANDARDS

Water-quality standards are usually based on one of two primary criteria: stream standards or effluent standards. *Stream standards* can be based on dilution requirements or the receiving-water quality based on a threshold value of specific pollutants or a beneficial use of the water. *Effluent standards* can be based on the concentration of pollutants which can be discharged or on the degree of treatment required.

Stream standards will usually be based on a system of classification of the water quality based on the intended use of the water. Table 1.2 generally shows typical classifications, the primary quality criteria, and the usual degree of treatment needed to meet these criteria.

Although stream standards are the most realistic in light of use of the assimilative capacity of the receiving water, they are difficult to administer and control in an expanding industrial and urban area. The equitable allocation of pollutional loads for a multiplicity of industrial and municipal complexes also poses political and economic difficulties. A stream standard based on minimum dissolved oxygen at low stream flow intuitively implies a minimum degree of treatment. One variation of stream standards is the specification of a maximum concentration of a pollutant (that is, the BOD) in the stream after mixing at a specified low flow condition.

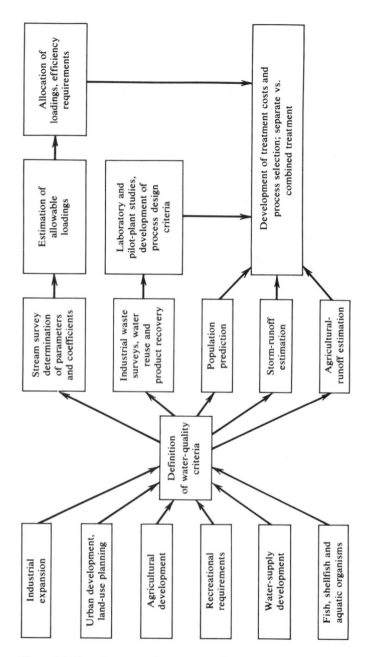

Figure 1.1 Development of a water-quality management program.

Table 1.2. Stream Classification for Water-Quality Criteria[a]

Class	Use	Quality Criteria	Required Treatment
A[b]	Water supply, recreation	Coliform bacteria, color, turbidity, pH, dissolved oxygen, toxic materials, taste- and odor-producing chemicals, temperature	Secondary (tertiary in some cases to meet criteria) plus disinfection
B[b]	Bathing, fish life, recreation	Coliform bacteria, pH, dissolved oxygen, toxic materials, color and turbidity (at high levels), temperature	Secondary plus disinfection
C	Industrial, agricultural, navigation, fish life	Dissolved oxygen, pH, floating and settleable solids, temperature	Primary and, in some cases, secondary
D	Navigation, cooling, water, etc.	Nuisance-free conditions, floating material, pH	Primary

[a] From [2, 3].
[b] May require nutrient (nitrogen and phosphorus removal).

It should be noted that maintenance of water quality and hence stream standards are not static but subject to change with the municipal and industrial environment. For example, as the carbonaceous organic load is removed by treatment, the detrimental effect of nitrification in the receiving water increases. Eutrophication may also become a serious problem in some cases. These considerations will require an upgrading of the required degree of treatment.

Effluent standards may be based on the maximum concentration of a pollutant (mg/1) or the maximum load (lb/day) discharged to a receiving water. These standards are usually related to a stream classification.

In some cases, the degree of treatment is specified (i.e, primary, secondary, etc.). The difficulty with this criterion is that industrial wastes do not follow the pollutant-concentration levels or treatment sequences as do municipal wastes. For example, primary and secondary treatment of municipal wastewaters will yield 35 percent and 85 to 90 percent removal of BOD, respectively, with effluent concentrations of about 130 and 20 mg/1, respectively. The concentration of pollutants in industrial wastes will depend not only on the type of industry but also on housekeeping practices, water reuse, and other factors; in many cases it will be many times that of municipal wastes. The same percentage treatment may still impose an excessive load on the receiving water. Industries with good housekeeping practices may also be penalized.

The Commonwealth of Pennsylvania has attempted to equalize this effect by establishing average waste characteristics for a number of industrial categories and then applying the degree of treatment required to the industry average waste loading. In this way, an industrial plant with effective water-conservation programs will require less treatment than one with poor housekeeping practices.

In several cases, such as shellfish areas and aquatic reserves, the usual water-quality parameters do not apply because they are nonspecific as to detrimental effects on aquatic life (for example, COD is an overall measure of organic content but does not differentiate between toxic and nontoxic organics). In these cases, a species diversity index* has been employed as related to either free-floating or benthic organisms. (The species diversity index is an indication of the overall condition of the aquatic environment. It is related to the number of species in the sample. The higher the species diversity index, the more productive is the aquatic system.)

The water-quality criteria for various industrial uses are summarized in Table 1.3. The surface-water-quality criteria for public water supplies have been summarized in Ref. 1. Color should not exceed 75 units and odors should be virtually absent. The water temperature should not exceed 85°F and should be limited to a 5°F temperature increase over that caused by ambient conditions. Alkalinity should be between 30 and 400 to 500 mg/l. Nitrate plus nitrite is limited to 10 mg/l. Total dissolved solids should not exceed 500 mg/l and the carbon–chloroform extract 0.2 mg/l.

For freshwater fish and aquatic life the total dissolved solids should not exceed the equivalent of 1500 mg/l of chlorides. The pH should be between 6 and 9 and the alkalinity in excess of 20 mg/l. The temperature should not exceed 5°F greater than normal ambient. For most fish life, the dissolved oxygen should be in excess of 5.0 mg/l.

Water for irrigation is limited in concentration of total dissolved solids, 500 to 1500$^{(l)}$; sodium content, chlorides, 100 to 350$^{(l)}$ mg/l; copper 0.1 to 1.0 mg/l; and pH 6.0 to 9.0$^{(l)}$ [(l) refers to the maximum limiting concentration].

References

1. *Report of the Committee on Water Quality Criteria,* U.S. Department of Interior, Federal Water Pollution Control Administration, Washington, D.C. April 1968.
2. J. E. McKee and H. W. Wolf, *Water Quality Criteria,* Report to the California Water Quality Control Board, SWPCB Publ. 3A, 2nd ed., 1963.

---

*Species diversity index, $K_D = (S - 1)/\log_{10} I$, in which S is the number of species and I the total number of individual organisms counted.

Table 1.3. Industrial Water-Quality Limits[a]

Industry	Turbidity	Color	Hardness	Temp., °F	pH	TDS	Fe and Mn	SS	Cl
Textiles	5	5	25		6.0–8.0[b]	100	0.1	5	
Dyeing	5	5	25		6.0–8.0[b]	100	0.1	5	
Scouring	5	5	25		3.0–10.0	100	0.1	5	
Bleaching	5	5	25		2.5–10.5[b]	100	0.1	5	
Pulp and paper									
Mechanical pulping	—	30	c	c	6–10	c	0.3	c	1000
Unbleached	—	30	100	c	6–10	c	1.0	10	200
Bleached	—	10	100	95	6–10	c	0.1	10	200
Organic chemicals	—	—	250	—	6.5–8.7	c	0.1	c	—
Petroleum	—	c	350	—	6–9	1000	1	10	300
Iron and steel	—	—	—	100	5–9	—	—	100[a]	—
Cannery	—	5	250	—	6.5–8.5	500	0.4	10	250
Tanning	—	5	150	—	6–8	—	0.3[b]	—	—
Brewing	10	—	—	—	6.5–7	500	0.1	—	—

[a]From [1]; units are mg/l unless otherwise specified.
[b]Varies with product.
[c]Not considered a problem of concentrations encountered.
[a]Settleable solids.

3. P. H. McGauhey, *Engineering Management of Water Quality*, McGraw-Hill, New York, 1968.
4. S. R. Weibel et al., "Urban Land Runoff as a Factor in Stream Pollution," *J. Water Pollution Control Federation*, **36,** No. 7, 914 (1964).
5. R. O. Sylvester, "Nutrient Content of Drainage Water from Forested, Urban and Agricultural Areas," *Trans. Sem. Algae Metropolitan Wastes,* United States Public Health Service, R. A. Taft Center, Cincinnati, Ohio, 1960.

2

Sewage and Industrial-Waste Characterization

The increased emphasis on water quality for multipurpose use has defined a number of parameters of special significance in municipal sewage and industrial wastewaters. These are

1. The BOD (biochemical oxygen demand), which defines the biodegradable organic content of the waste.

2. The COD (chemical oxygen demand), which measures the total organic content, both degradable and refractory.

3. Suspended and volatile suspended solids.

4. Total solids.

5. pH: alkalinity and acidity.

6. Nitrogen and phosphorus.

7. Heavy metals and inorganic solids.

Certain industrial wastes will contain parameters of special significance, such as phenol or cyanide.

The significance of these parameters with respect to water-quality management is listed in Table 2.1.

ESTIMATING THE ORGANIC CONTENT OF WASTEWATERS

The organic content of a waste can be estimated by each of three tests, although considerable caution should be exercised in interpreting the results. The BOD test will measure the

Table 2.1   Undesirable Characteristics of Industrial Wastes

1 Soluble organics: dissolved oxygen depletion in streams and estuaries; discharge relative to assimilative capacity of water body or by effluent standard.

2 Soluble organics that result in tastes and odors in water supplies, e.g. phenol.

3 Toxic materials and heavy metal ions; e.g., cyanide, Cu, and Zn: usually rigid standards as to discharge of such materials.

4 Color and turbidity: esthetically undesirable; imposes increased loads on water-treatment plants; example: color from pulp and paper mills.

5 Nutrients (nitrogen and phosphorus): enhance eutropication of lakes and ponded areas; critical in recreational areas.

6 Refractory materials, e.g., ABS: results in foaming in streams.

7 Oil and floating material: regulations usually require complete removal; esthetically undesirable.

8 Acids and alkalis: neutralization required in most regulatory codes.

9 Substances resulting in atmospheric odors, e.g., sulfides from tanneries.

10 Suspended solids: results in sludge banks in streams.

11 Temperature: thermal pollution resulting in depletion of dissolved oxygen (lowering of saturation value).

biodegradable organic carbon and, under certain conditions, the oxidizable nitrogen present in the waste. The COD will measure the total organic carbon with the exception of certain aromatics, such as benzene, which are not completely oxidized in the reaction. The COD test is an oxidation–reduction reaction, so other reduced substances, such as sulfides, sulfites, and ferrous iron, will also be oxidized and reported as COD. The TOC test measures all carbon as CO_2, and hence the inorganic carbon (CO_2, HCO_3, and so on) present in the wastewater must be removed prior to the analysis or corrected for in the calculation. It should be emphasized that considerable caution must be exercised in interpreting the test results and in correlating the results of one test with another.

BIOCHEMICAL OXYGEN DEMAND

The BOD is conventionally reported as the 5-day value and is defined as the amount of oxygen required by living organisms engaged in the utilization and stabilization of the organic matter present in the wastewater. The standard test involves seeding with sewage, river water, or effluent, and incubating at 20°C.

Reaction in the BOD Bottle   The reaction in the BOD bottle is the same as in all aerobic reactions and occurs in two separate and distinct phases, as shown in Fig. 2.1. Initially, the organic matter present in the wastewater is utilized by the seed microorganisms for energy and growth. This results in a utilization of oxygen and the growth of new microorganisms. When the organics originally present in the wastewater are removed, the organisms present continue to use oxygen for autooxidation or endogenous metabolism of their cellular mass. When the cell mass is completely oxidized, only a nonbiodegradable cellular residue remains and the reaction is complete. This is defined as the ultimate BOD. The oxidation of BOD is therefore a two-phase reaction.

The removal and oxidation of the organics present in the wastewater is usually complete in 18 to 36 hr (phase 1). The total oxidation of the cell mass will take more than 20 days (phase 2). The rate of reaction during the first or assimilation phase is 10 to 20 times the rate of endogenous oxidation. These relative rates are shown in Fig. 2.1.

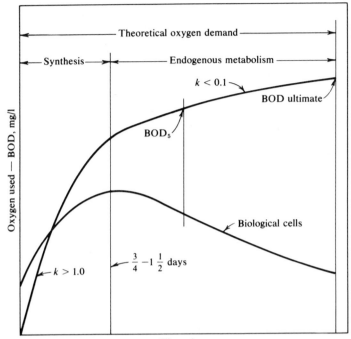

Figure 2.1 Reactions occurring in the BOD bottle.

Formulation of the BOD   The BOD has been class-
ically formulated as a continuous first-order reaction of the
form

$$y = L_0(1 - 10^{-kt}) \tag{1}$$

where y = amount of oxygen consumed or BOD after any
     time t
   L_0 = ultimate BOD or the total amount of oxygen
     consumed in the reaction.
   k = average reaction-rate constant
   t = time of incubation, days

It should be emphasized that because the reaction in the BOD bottle consists of two separate and distinct phases with an order-of-magnitude difference in reaction rate, the mean k in equation (1) will vary markedly, depending on the quantity and nature of organics present in the wastewater. For example, consider a waste that is treated through an activated sludge process. In the raw waste a rapid consumption of oxygen occurs (phase 1, Fig. 2.1) followed by the slower endogenous rate, yielding a high mean k in equation (1).

In the treated effluent most of the organics originally present in the wastewater have been removed, so most of the oxygen is consumed at the slower endogenous rate. This, in turn, yields a lower mean k as compared to the untreated wastewater.

Typical values of the mean rate constant k are summarized in Table 2.2. It can be seen from Table 2.2 that when compar-

Table 2.2.   Average BOD Rate Constants at 20°C

Substance	k_{10}
Untreated wastewater	0.15–0.28
High-rate filters and anaerobic contact	0.12–0.22
High-degree biotreatment effluent	0.06–0.10
Rivers with low pollution	0.04–0.08

ing wastes or treatment-process efficiencies or when estimating the ultimate BOD, cognizance must be taken of the reaction rate k. Typical BOD curves for a raw waste and a treated effluent are shown in Fig. 2.2.

Both k and L_0 are unknown in the BOD reaction [equation (1)], so an indirect calculation must be employed. Several procedures have been developed for this, three of which are summarized below.

1. Method of moments (Moore et al.) [1]. From a smoothed curve of the data, tabulation of t, y, and ty is made for a given numerical sequence of days (that is, 1, 2, 3). From prepared

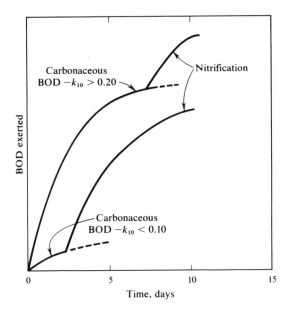

Figure 2.2 Comparison of BOD curves for a raw waste and a treated effluent.

charts, k is determined from the ratio $\Sigma y/\Sigma ty$ and L_0 from $\Sigma y/L_0$. To minimize error, the data should be plotted, a smooth curve drawn through the points, and k and L computed from Fig. 2.3.

2. Log-difference method. The BOD equation is expressed

$$y = L_0(1-10^{-kt}) \tag{1}$$

Differentiating equation (1) yields

$$\frac{dy}{dt} = r = L_0 k e^{-kt} \tag{2}$$

in which r is the rate of oxygen utilization with time. Equation (2) is a semilogarithmic plot of the form

$$\log r = \log L_0 k - kt$$

From the corrected values of y, differences are calculated and tabulated. The differences are plotted vs. time on semilog

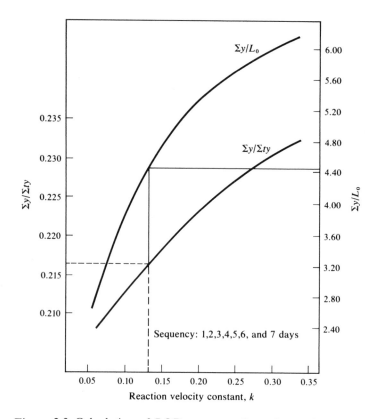

Figure 2.3 Calculation of BOD constants from the method of moments. (After Moore et al. [1].)

paper. From this plot k and L are computed. This method can take into account nonconformance with first-order kinetics.

3. Graphical method (Thomas)[2]. $(t/y)^{1/3}$ is plotted as the ordinate vs. t as the abscissa. From the plot,

$$k = 2.61 \frac{b}{a}, \quad L_0 = \frac{1}{2.3} ka^3$$

where b is the slope of the line and a the intercept.

An example calculation using the log-difference method is given.

1. Plot y vs. t on cartesian-coordinate paper. Draw a curve of best fit through the observed points, including lags and plateaus, if any, to get a picture of the deoxygenation curve. If a majority of observed values describe a smooth progression, those that fail to fit may be considered erratic and curve values may be used for subsequent calculations. Lags should be eliminated by curve fitting and taking the observed points after the lag termination.

2. Plot the daily difference, corrected if necessary, on semilog paper with time on the linear scale and the daily difference on the log scale. The differences are conventionally plotted as $\frac{1}{2}$, $1\frac{1}{2}$, $2\frac{1}{2}$ days, and so on, to illustrate intervals rather than points (Table 2.3).

Table 2.3.   Daily Differences

t	y	Daily difference from plot
0	0	—
1	7.3	7.3
2	12.8	5.5
3	16.0	3.3
4	20.1	4.3
5	22.5	2.4
6	23.8	1.3
7	25.3	1.5

3. From Fig. 2.4 the value at the zero intercept of the daily difference slope is equal to L_0k. Other intercepts are a function of L_t. The number representing L_0k at zero time $= 8.3$.

$$k_{10} = \frac{\log (8.3/1.3)}{7} = 0.115$$

$$\text{intercept} = L_0k$$
$$8.3 = 2.3 \cdot 0.115 \cdot L_0$$
$$L_0 = 31 \text{ mg/l}$$

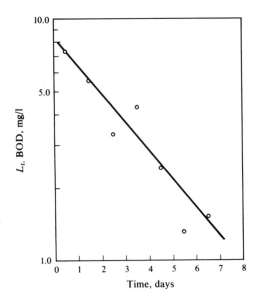

Figure 2.4   Calculation of k and L_0 by the difference method.

The ratio between the 5-day and the ultimate BOD as related to the reaction rate k is shown in Fig. 2.5.

Temperature Effects on the BOD   The BOD reaction rate, k,is directly affected by temperature. L_0 is slightly affected because oxidizability increases with temperature. The relationship, derived from van't Hoff's law, is

$$k_t = k_{20}\Theta^{(T-20)}$$
$\Theta = 1.047$ (Phelps) (inaccurate at low temperatures)
$= 1.056$ (20–30°C) (Schroepfer[3])
$= 1.135$ (4–20°C) (Schroepfer[3])

Factors Affecting the BOD   Several factors will affect the BOD test and should be considered, particularly when dealing with industrial wastes.

Seed   The acclimated organisms present below industrial outfalls provide an excellent source of seed for the BOD determination. With few exceptions, carefully selected seed

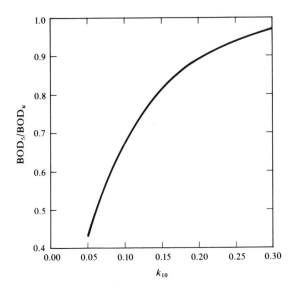

Figure 2.5 Relationship between k_{10} and BOD_5/BOD_u.

from the receiving stream will yield the highest BOD values. When streamwater is used for seed, nitrification difficulties may increase.

Occasionally it is necessary to artificially develop a microbial culture that will oxidize the industrial waste. An acclimatized heterogenous microbial culture may be developed by starting with settled domestic wastewater containing a large variety of organisms to which a small amount of the industrial effluent is added. The amount of waste added is increased until a culture develops that is adapted to the waste. The mixture of domestic wastewater and the industrial waste is aerated by bubbling air continuously through the liquid. A noticeable increase in the cloudiness or turbidity of the aerating mixture generally indicates an acclimated culture. If a DO probe is available, the oxygen uptake can be evaluated daily to determine when an acclimated culture has developed. The effect of acclimation is shown in Fig. 2.6.

The amount of seed required to produce a normal rate of oxidation must be determined experimentally. The most

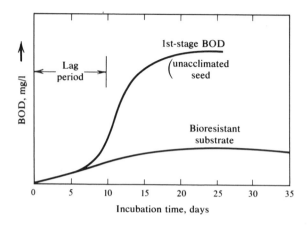

Figure 2.6 Effect of acclimation on BOD exertion.

frequent error is the use of insufficient seed. The effects of seed concentration on the BOD are illustrated in Fig. 2.7.

Large numbers of algae in streamwater that is used for dilution water may produce significant changes in the oxygen content. When stream samples containing algae are incubated in the dark, the algae survive for a time. Short-term BOD determinations may show the influence of oxygen production by the algae. After prolonged lack of light, the algae die and the algal cells contribute to the total organic content of the sample and increase the BOD; therefore, samples incubated in the dark may not be representative of the deoxygenation process in the stream, because the benefits of photosynthesis are lacking. On the other hand, samples incubated in the light, under conditions of continual photosynthesis, will yield low BOD values. The influence of algae in the BOD test is difficult to evaluate, and extreme care should be taken when streamwater that contains large numbers of algae is used for dilution water.

Toxicity .Various chemical compounds are toxic to microorganisms. At high concentrations the substances will kill the microbes, and at sublethal concentrations the activity of microbes is reduced. The effects of heavy metals on the

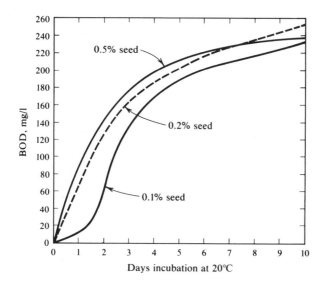

Figure 2.7 Effect of seed concentration on BOD exertion.

BOD are illustrated in Fig. 2.8. Toxicity is usually evidenced by an increase in BOD with increasing dilution.

Nitrification   The oxidation process described by the BOD equation $y = L_0(1 - 10^{-kt})$ represents the oxidation of carbonaceous matter:

$$C_x H_y O_z \longrightarrow CO_2 + H_2O$$

The oxidation of nitrogenous material may be shown as

$$NH_3 \longrightarrow NO_2 \longrightarrow NO_3$$

The rate constant is usually less than in the case of the carbonaceous matter.

Under some circumstances these two oxidations can proceed simultaneously, and the resultant BOD curve will be a composite of the two reactions. Normally, however, the nitrification will not begin until the carbonaceous demand has

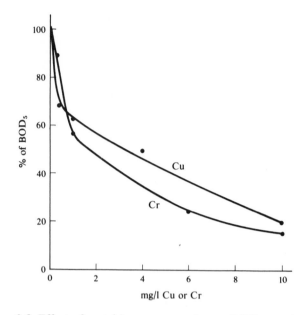

Figure 2.8 Effect of metal-ion concentration on BOD exertion. (After Morgan and Lackey [4].)

been partially satisfied, yielding a curve similar to Fig. 2.9. Mathematically the reactions can be described by

$$y = L_0(1 - 10^{-k_1 t}) + L_N(1 - 10^{-k_2 t})$$

where L_o = ultimate carbonaceous demand
$\quad\quad L_N$ = ultimate nitrogenous demand
$\quad\quad k_1$ = rate constant for carbonaceous demand
$\quad\quad k_2$ = velocity constant for nitrogenous demand

Nitrification occurs most often in effluents that have undergone partial oxidation of the waste components. Nitrification represents a demand on the oxygen resources of the receiving stream; therefore, it should be recognized as part of the total demand of the waste.

Nitrification can be eliminated by pasteurizing and reseeding or by the addition of methylene blue or thiourea. The rate of nitrification can be determined by a parallel set of BOD samples, one with and one without nitrification suppression.

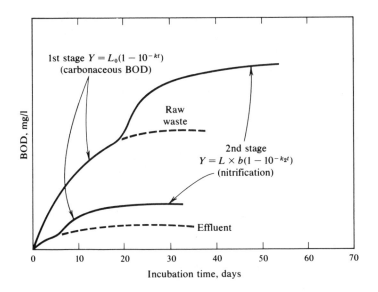

Figure 2.9 First and second stage BOD in raw waste and treated effluent.

Other Procedures for the Evaluation of the BOD Short-term TOD test (Busch)[5]: The oxygen consumed in 24 to 28 hr yields a plateau representative of oxygen consumed for respiration and synthesis. This is equal to about 40 percent of total oxygen demand. The concentration of cells produced at the plateau is measured. The oxygen due to endogenous respirations equals the cells times 1.42. The TOD is therefore equal to the oxygen consumed up to the plateau plus the oxygen consumed for endogenous respiration. The drawbacks to this method are the difficulty in obtaining an accurate weight of cells produced and the inability to accurately detect the plateau from one sample.

CHEMICAL OXYGEN DEMAND

The procedure for chemical oxygen demand as given by *Standard Methods*[10] employs potassium dichromate with reflux. Silver sulfate (Ag_2SO_4) is added as a catalyst, and when chlorides are present, Hg_2SO_4 can be added to complex the chlorides and eliminate the need for a chloride correction. All organic compounds are not oxidizable chemically by the dichromate procedure. Sugars, branched-chain aliphatics, and substituted benzene rings are completely oxidized with little or no difficulty. However, benzene, pyridine, and toluene are not oxidized by this method. Other compounds (straight-chain acids, alcohols, and amino acids) can be completely oxidized in the presence of the silver sulfate catalyst.

Chlorides interfere with the COD analysis; the interference is eliminated by using mercuric sulfate, which complexes the chlorides. The theoretical COD of organic compounds may be calculated if the reaction is known. The oxidation of 1000 mg of phenol is used as an example:

$$C_6H_5OH + 7O_2 \longrightarrow 6CO_2 + 3H_2O$$

$$\text{theoretical COD} = \frac{(1000)(224)}{94} = 2383 \text{ mg}$$

When the constituents of a waste are known, it is at times helpful to estimate the theoretical COD to determine the yield of the dichromate refluxing procedure. In this manner the ability of the COD test to provide a representative estimate of the strength of the waste is established. The preparation of the required reagents, the required apparatus, and the procedure are discussed in detail in *Standard Methods*[*10*].

A short-term rapid chemical oxygen demand test[*6*] can be used for routine plant control or for comprehensive waste surveys. This test employs a short oxidation period, so some of the more complex organics may be incompletely oxidized in this test. It is, therefore, desirable to compare the results of the short-term test with the *Standard Methods* test for a particular wastewater. Some comparative results reported by Jeris[*6*] are summarized in Table 2.4.

TOTAL ORGANIC CARBON

Total organic carbon (TOC) can be measured either manually or automatically.

Table 2.4.   Rapid COD Test Results[a]

Compound	COD, mg/l		
	Theoretical[b]	Rapid Method	Standard Method
Glucose	1299	1264	1258
Pyridine	2185	36	0
Ethanol	821	715	777
Acetic acid	772	729	727
Glycine	1243	250	505
Phenylalanine	1390	1330	1347
Nutrient broth		492	536
Activated sludge		2150	2266
Digested sludge		1409	1441
Raw sewage		183	202
Settled sewage		172	183

[a] After Jeris [6].
[b] Computed using the weight of compound.

The manual or wet oxidation method involves oxidation of the sample in a solution of chromic acid, potassium iodate phosphoric acid, and fuming sulfuric acid. The combustion products are swept through a Pregl-type combustion tube and the resulting carbon dioxide collected and weighed in an absorption train.

Carbon Analyzer The recently developed carbon analyzer utilizes the concept of complete combustion of all organic matter to carbon dioxide and water; then the gas stream is allowed to pass through an infrared analyzer sensitized for carbon dioxide and the response is recorded on a strip chart.

The carbon analyzer is shown in Fig. 2.10. A microsample of the wastewater to be analyzed is injected into the catalytic combustion tube, which is maintained at a temperature of 900 to 1000°C. The sample is vaporized and the carbonaceous material is completely oxidized in the presence of a cobalt catalyst and the pure oxygen carrier gas. The oxygen flow carries the carbon dioxide and steam out of the furnace; the steam

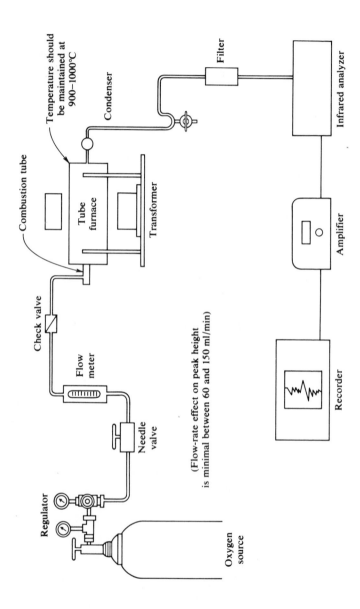

is condensed; and the remaining carbon dioxide, oxygen, and water vapor mixture enters the infrared analyzer. As the amount of carbon dioxide is directly proportional to the initial sample carbon concentration, the peak-height response can be compared to a calibration curve and the value determined. This value includes carbon in the form of carbonates or bicarbonates which may have been present. To measure only organic carbon, all inorganic carbon must first be removed from the sample. This is accomplished by acidifying the sample and purging with nitrogen gas prior to injection. The sample is measured for carbon before and after the acidification–purging step, any reduction being attributed to the loss of inorganic carbon and volatile organic compounds. The volatile organic fraction, if present, must be measured independently, and this is accomplished by using a closed-system diffusion-cell method[7].

A recent modified carbon analyzer has facilitated the measurement of total organic carbon, as shown in Fig. 2.ll. A dual-combustion tube system allows direct differentiation between organic and inorganic carbon, making the preliminary acidification and scrubbing step unnecessary. A microsample is first injected into the high-temperature combustion tube and the total carbon concentration determined by the method shown in Fig. 2.10. An identical volume of sample is then injected into the low-temperature combustion tube, which contains quartz chips wetted with phosphoric acid. The operating temperature of 150°C is below the value at which organic matter is oxidized, but in the acid environment all carbonates are removed in the form of carbon dioxide. This gas then flows through the infrared analyzer and is measured as inorganic carbon. The total organic carbon concentration is taken as the difference between the total carbon and total inorganic carbon values. An optional air-purification unit is available to provide a suitable carrier gas should a bottled source not be desirable.

Some interference in the infrared-energy-absorption pattern is possible if anions such as NO_3^-, Cl^-, SO_4^{--}, and PO_4^{---} are present in excess of 10,000 mg/l. Industrial wastewaters containing such concentrations should therefore be diluted with carbon dioxide–free water prior to analysis.

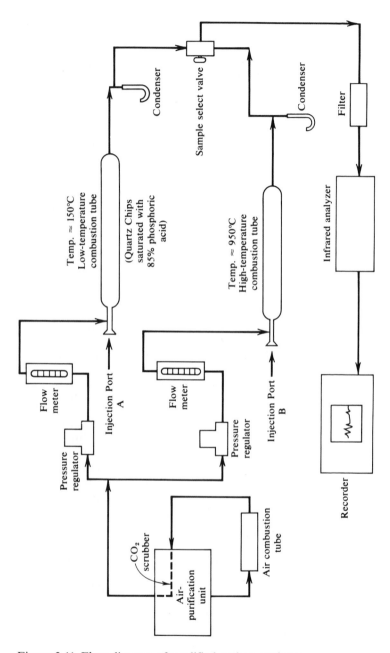

Figure 2.11 Flow diagram of modified carbon analyzer.

RELATIONSHIPS AMONG BOD, COD, AND TOC

When considering routine plant control or investigational programs, the BOD is not a useful test because of the long incubation time required to obtain a meaningful result. It is therefore important to develop correlations among BOD, COD, and TOC.

Let us first consider a completely biodegradable substance such as glucose. The ultimate BOD will measure about 90 percent of the theoretical oxygen demand. (Approximately 10 percent of the original organics end up as nonbiodegradable cellular residue and hence is not measured in the BOD.) The COD will measure the theoretical oxygen demand. Therefore, for these substrates

$$COD = \frac{BOD_u}{0.9} = TOD$$

ORGANIC CARBON–OXYGEN DEMAND RELATIONSHIP

In attempting to correlate BOD or COD of a wastewater with TOC, one should recognize those factors which may discredit the correlation. These include the following:

1. A portion of the COD of many industrial wastes is attributed to the dichromate oxidation of ferrous iron, nitrogen, sulfites, sulfides, and other oxygen-consuming inorganics.

2. The BOD and COD tests do not include many organic compounds that are partially or totally resistant to biochemical or dichromate oxidation. However, all the organic carbon in these compounds is recovered in the TOC analysis.

3. The BOD test is susceptible to variables that include seed acclimation, dilution, temperature, pH, and toxic substances.

One would expect the stoichiometric COD/TOC ratio of a wastewater to approximate the molecular ratio of oxygen to carbon ($\frac{32}{12} = 2.66$). Theoretically the ratio limits would range from zero, when the organic material is resistant to dichromate oxidation, to 5.33 for methane or slightly higher when inorganic reducing agents are present.

Reported BOD, COD, and TOC values for several organic compounds are given in Table 2.5 and industrial wastewaters

Table 2.5.   Relationship between COD and TOC for Organic Compounds[a]

Substance	COD/TOC (Calculated)	COD/TOC (Measured)
Acetone	3.56	2.44
Ethanol	4.00	3.35
Phenol	3.12	2.96
Benzene	3.34	0.84
Pyridine	3.33	Nil
Salicylic acid	2.86	2.83
Methanol	4.00	3.89
Benzoic acid	2.86	2.90
Sucrose	2.67	2.44

[a] After Ford [8].

are listed in Table 2.6. The COD/TOC ratio varies from 1.75 to 6.65[8].

Although it has been difficult to correlate BOD with TOC for industrial wastes, relatively good correlation has been obtained for domestic wastewaters. This is reasonable when

Table 2.6.   Industrial Waste-Oxygen Demand and Organic Carbon[a]

Type of Waste	BOD_5, mg/l	COD, mg/l	TOC, mg/l	BOD/TOC	COD/TOC
Chemical[b]	—	4,260	640	—	6.65
Chemical	24,000	41,300	9,500	2.53	4.35
Chemical	850	1,900	580	1.47	3.28
Chemical	700	1,400	450	1.55	3.12
Olefin processing	—	321	133	—	2.40
Chemical	—	350,000	160,000	—	2.19
Synthetic rubber	—	192	110	—	1.75

[a] After Ford [8].
[b] High concentration of sulfides and thiosulfates.

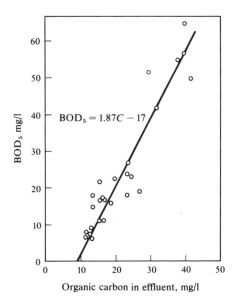

Figure 2.12 Relationship between BOD_5 and organic carbon in domestic sewage biologically treated effluent.

one considers the type and consistency of the waste constituents. A 5-day BOD–TOC correlation for sewage has been reported by several investigators. A ratio of 1.87 has been reported by Wurhmann[9]. Mohlman and Edwards have reported a range of 1.35-2.62 for raw domestic waste. The calculated relationship between 5-day BOD and TOC (see Fig. 2.12) is

$$\frac{BOD_5}{TOC} = \frac{O_2}{C} = \frac{32}{12}(0.90)(0.77) = 1.85$$

where:

1. The ultimate BOD will exert approximately 90 percent of the theoretical oxygen demand.

2. The 5-day BOD is 77 percent of the ultimate BOD for domestic wastes.

A decrease in the COD/TOC and BOD_5/COD ratios has been observed during the biological oxidation of both munici-

Table 2.7.   Variation of COD/TOC and BOD₅/TOC through Biological Treatment

Waste	COD/TOC		BOD₅/TOC	
	Raw	Effluent	Raw	Effluent
Domestic	4.15	2.20	1.62	0.47
Chemical	3.54	2.29	—	—
Refinery—chemical	5.40	2.15	2.75	0.43
Petrochemical	2.70	1.85		

pal and industrial wastewaters, as shown in Table 2.7 and in Fig. 2.13. This can be attributed to

1. The presence of inorganic reducing substances that would be oxidized in the biological process, thereby reducing the COD/TOC ratio.

2. Intermediate compounds may be formed during the biological process without significant conversion of organic matter to carbon dioxide. A reduction in COD may not be accompanied by a reduction in TOC.

3. The BOD reaction-rate constant k will be greater than 0.15 in the raw waste and less than 0.1 in the treated effluent. The BOD_5/BOD_u and hence the BOD_5/COD ratio depends on this rate. This is at least in part responsible for the reduction in the BOD_5/COD or BOD_5/TOC during biological oxidation.

4. The concentration of nonremovable refractory materials will account for a larger portion of the COD in the effluent than in the raw waste, thereby lowering the BOD_5/COD or the BOD_5/TOC ratio.

References
1. E. W. Moore et al., "Simplified Method for Analysis of BOD Data," *Sewage Ind. Wastes,* **22,** 1343 (1950).
2. H. A. Thomas, "Graphical Determination of BOD Curve Constants," *Water Sewage Works,* **97,** 123 (1950).
3. G. S. Schroepfer et al., in *Advances in Water Pollution Control,* Vol. I, Pergamon Press, Oxford, 1964.

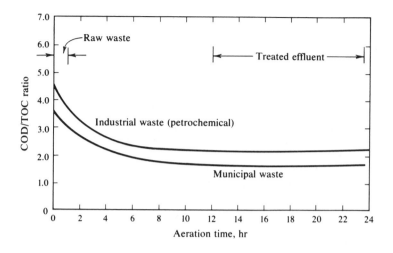

Figure 2.13 COD/TOC ratio at various stages of biological oxidation.

4. G. B. Morgan and J. B. Lackey, "BOD Determinations in Wastes Containing Cheleated Copper or Chromium," *Sewage Ind. Wastes,* **30,**No. 3, 283 (1958).

5. A. W. Busch, "BOD Progression in Soluble Substrates," *Proc. l5th Ind. Waste Conf.,* Purdue University, Lafayette, Ind., 196l.

6. J. S. Jeris, "A Rapid COD Test," *Water Wastes Eng.,* **5,** No. 4,89 (1967).

7. R. B. Schaffer et al., "Application of a Carbon Analyzer in Waste Treatment," *J. Water Pollution Control Federation,***37,** No. 11, 1545 (1965).

8. D. L. Ford, "Application of the Total Carbon Analyzer for Industrial Wastewater Evaluation," *Proc. 23rd Ind. Waste Conf.,* Purdue University, Lafayette, Ind., 1968.

9. K. Wuhrmann, *Hauptwirkungen und Wechsel Wirk Kungen einiger Betriebsparameter im Belebschlamm-system Ergebnissemehrjahriger,* Grossversuche Verlag, Zurich, 1964.

10. APHA, AWWA, etc., *Standard Methods for the Examination of Water and Wastewater,* 12th ed., Amer. Public Health Ass., Inc., 1965.

3

Analysis of Pollutional Effects in Natural Waters

The movement and reactions of waste materials through streams, lakes, and estuaries is a resultant of hydrodynamic transport and biological and chemical reactions by the biota, suspended materials, plant growths, and bottom sediments. These relationships can be expressed by a mathematical model that reflects the various inputs and outputs in the aquatic system. Considering the oxygen balance, the general relationships for the oxygen sag curve are

$$\frac{\partial C}{\partial t} = \epsilon \frac{\partial^2 C}{\partial X^2} - U \frac{\partial C}{\partial X} \pm \Sigma S \tag{1}$$

where C = concentration of dissolved oxygen
t = time at a stationary point
U = velocity of flow in the X direction
ϵ = turbulent diffusion coefficient
S = sources and sinks of oxygen
X = distance downstream

Equation (1) assumes that the concentration of any characteristic is uniform over the stream cross-reaction and that the area is uniform with distance. If this is not the case, equation (1) must be suitably modified.

The sources and sinks of oxygen can be listed as follows:
Sources of oxygen:
1. Quantity in incoming or tributary flow.
2. Photosynthesis.
3. Reaeration.
Sinks of oxygen:
1. Biological oxidation of carbonaceous organic matter.
2. Biological oxidation of nitrogenous organic matter.
3. Benthal decomposition of bottom deposits.
4. Respiration of aquatic plants.
5. Immediate chemical oxygen demand.

SOURCES OF OXYGEN

1. The quantity of oxygen in the incoming or tributary flow is considered as an initial condition in equation (1). The dissolved oxygen present in waste discharges should also be considered if the waste flow is large relative to the stream flow.

2. The degree of photosynthesis depends upon sunlight, temperature, mass of algae, and available nutrients and will exhibit a diurnal variation, as shown in Fig. 3.1.

3. Oxygen will be added to the water body by the process of natural reaeration. Reaeration is primarily related to the degree of turbulence and natural mixing in the water body (high in sections of rapids; low in impounded areas). The oxygen transfer from air to water can be defined as

$$N = K_L A (C_s - C_L) \tag{2}$$

where N = lb of O_2/hr
$\quad K_L = O_2$ transfer coefficient
$\quad A$ = surface area
$\quad C_s = O_2$ saturation concentration
$\quad C_L = O_2$ concentration
and, in concentration units,

$$\frac{dC}{dt} = \frac{K_L A}{V}(C_s - C_L) \tag{3}$$

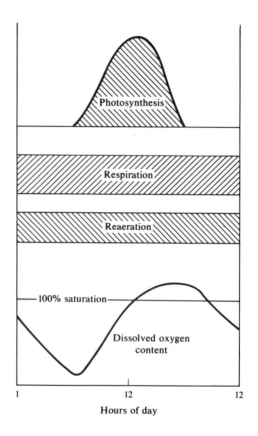

Figure 3.1 Diurnal effects in a stream due to photosynthesis.

for a stream $A/V = 1/H$ and $(C_s - C_L)$ = oxygen deficit (D) and equation (3) reduces to

$$\frac{dC}{dt} = \left(\frac{K_L}{H}\right) D = K_2 D \qquad (4)$$

in which K_2 is the reaeration coefficient. It should be noted that $A/V = 1/H$ applies only to a quiescent surface and would increase for a turbulent surface. Dobbins[7] estimates $A/V = 1.5/H$ for a highly turbulent stream surface.

The reaeration coefficient K_2 has usually been defined by a relationship of the type

$$K_2 = \frac{CV^n}{H^m} \tag{5}$$

in which V is the average stream velocity and H the average stream depth. The coefficient is a function of stream characteristics. Organics and surface-active agents present in the streamwater will affect K_2 and this effect will be reflected in the constant C.

The exponents m and n also relate to stream conditions. Reported values of the coefficients are summarized in Table 3.1.

Table 3.1. Summary of Coefficients for the Reaeration Equation[a]

$$K_2 = C\frac{V^n}{H^m}$$

C	n	m	Reference
12.9	$\frac{1}{2}$	$\frac{3}{2}$	[2]
5.0	1.0	$\frac{5}{3}$	[8]
3.3	1.0	$\frac{4}{3}$	[12]

[a]V is in fps and H in ft.

SINKS OF OXYGEN

1. In the biological oxidation of carbonaceous organic matter, the rate of removal, K_r, is related to amount of unstabilized organics present:

$$L = L_0 e^{-K_r x/U} = L_0 e^{-K_r t} \tag{6}$$

where L is concentration of organics present at time t, L_0 is concentration of organics present at time zero, and K_r relates to the removal of organics by all mechanisms: sedimentation, oxidation, and volatilization. For oxidation alone, as might result from a soluble organic waste, equation (6) is expressed

$$L = L_0 e^{-K_1 x/U} = L_0 e^{-K_1 t} \tag{7}$$

K_r will be considerably greater than K_1 when suspended or volatile organics are present (see Fig. 3.2). The rate of removal

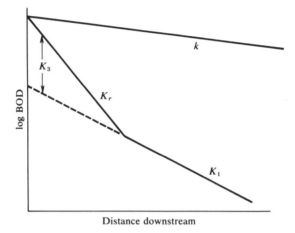

Figure 3.2 Deoxygenation relationship in a stream.

of organics by mechanisms other than oxidation has been defined as K_3. The rate of oxidation of the organics in the BOD bottle is defined as k. k is usually less than K_1 because longitudinal mixing, the presence of bottom growths, and suspended biological solids will increase the reaction rate. (The concentration of seed organisms will also usually be higher in a stream.) Values of k will depend upon the characteristics of the waste, decreasing with treatment or removal of readily oxidizable organics. The reported range of values to be expected is given in Table 3.2. By contrast, K_1 may have values in excess of 20 per day. Bosko[11] has related k to K_1 through the characteristics of the stream:

$$K_1 = k + \frac{V}{H} \eta \qquad (8)$$

in which V is the velocity of stream flow, H the depth of the stream, and η a coefficient of bed activity which may vary from 0.1 for stagnant or deep waters to 0.6 or higher for rapidly flow-

Table 3.2. Values of k

Substance	$k_{(e)}$/day
Raw sewage, high-rate treatment effluent	0.35–0.60
High-degree biological effluent	0.10–0.15
Rivers with low pollution	0.10–0.12

ing streams. These various reaction rates are compared in Fig. 3.2.

2. When unoxidized nitrogen is present in the wastewater, nitrification will result with time of passage or distance downstream. The rate of nitrification will be less when untreated wastes are discharged to the stream and the concentration of nitrifying organisms is low, and will increase in the presence of well-oxidized effluents with a high seed of nitrifying organisms.

The reactions are

Oxidation of ammonia to nitrite by *Nitrosomonas*:

$$2NH_4^+ + 3O_2 \longrightarrow 2NO_2^- + 2H_2O + 4H^+$$

Oxidation of nitrite to nitrate by *Nitrobacter*:

$$2NO_2^- + O_2 \longrightarrow 2NO_3^-$$

Nitrifying organisms are sensitive to pH and function best over a pH range 7.5 to 8.0. The rate of nitrification decreases rapidly at dissolved oxygen levels below 2.0 to 2.5 mg/l, so that at low oxygen levels in the water body little or no nitrification will occur. Denitrification has been observed to occur in stretches of zero or near-zero oxygen concentration. The kinetics of nitrification can be expressed as an autocatalytic reaction,

$$\frac{dC}{dt} = \pm KC(N - C) \tag{9}$$

where C = concentration of ammonia or nitrate
 N = concentration initially present
 K = reaction coefficient

which integrates to

$$C = \frac{N}{1 \pm e^{KN(\alpha - t)}} \tag{9a}$$

in which α = time for half completion of reaction. The equation can be linearized,

$$\frac{N - C}{C} = e^{KN(\alpha - t)} \tag{9b}$$

and plotted as log $[(N - C)/C]$ vs. t to evaluate the reaction rate K. Data can also be correlated as two forms of a first-order reaction:

$$\frac{dC}{dt} = -K(N - C) \tag{10}$$

and

$$\frac{dC}{dt} = -KC \tag{10a}$$

at $t = 0$, $C = N/2$ for both curves, and integration yields

$$C = N\left(1 - \frac{e^{-Kt}}{2}\right) \tag{10b}$$

and

$$C = \frac{N}{2} e^{-Kt} \tag{10c}$$

The temperature effect on the rate of nitrification in natural waters has been expressed[13] as
Nitrosomonas (river water):

$$K = 0.47 \cdot 1.106^{(T - 15)}$$

Nitrobacter (river water):

$$K = 0.79 \cdot 1.072^{(T - 15)}$$

The usual operational equation for a stream may be considered:

$$N = N_0 e^{-K_n x/U} = N_0 e^{-K_n t} \tag{11}$$

It should be recognized that because the growth rate of the nitrifying organisms is considerably lower than the carbonaceous organisms, oxygen depletion through nitrification will lag the deoxygenation from carbonaceous organics. When secondary-sewage-treatment plants are installed, the quantity of carbonaceous organics to be removed is greatly reduced, but much larger numbers of nitrifying organisms are present in the stream. Under these conditions nitrification is more rapid and may exert a significant oxygen demand.

3. In areas where bottom deposits occur, oxygen will be used by diffusion into the upper layers of the deposit (the rate will increase in the presence of worms, which increases the porosity of the deposit) and from the diffusion of organic products of anaerobic degradation into the flowing streamwater, which will increase the soluble organic oxygen demand of the water as shown in Fig. 3.3 (see also Table 3.3). The oxygen consumption has been observed to increase somewhat with dissolved oxygen concentration. Invertebrates living in muds, such as midge larvae, increase the interchange between the mud and the overlying water. No effect is reported on oxygen consumption with sediment depth.

Table 3.3. Characteristics of River Deposits

	Mean	Range
Dry solids, % of sediment	21	10–40
Volatile solids,		
% of dry	20	11–27
g/m²	460	150–2000
Organic nitrogen, % of volatile	3.5	1.3–5.1

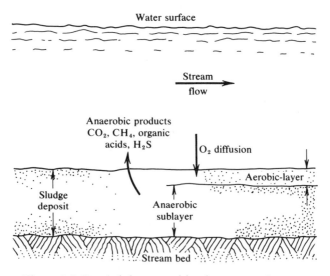

Figure 3.3 Benthal decomposition in a stream bottom.

4. Oxygen will be removed from the water body by the respiration of aquatic plants. (Oxygen may also be contributed by photosynthesis.) The oxygen loss is usually expressed on a weight basis or on an area basis (g of O_2/m^2). Oxygen will also be consumed by the development of sewage fungus (*Sphaerotilus*) when available carbohydrates are present in the water. The Water Pollution Research Laboratory[14] showed *Sphaerotilus* growths of 18 g/m² at 10°C and 3 g/m² at 20°C. They showed that maximum growth occurred at a stream velocity of 0.45 to 0.9 fps (0.14 to 0.27 m/sec). Higher velocities promote scour, and lower velocities have insufficient turbulence for distribution of the nutrient supply.

5. Many industrial wastes contain chemicals that will exert an immediate oxygen demand, such as sulfites. This will usually be immediately apparent in the oxygen sag curve and the BOD curve in the stream.

Mathematical Development of the BOD–Oxygen Sag Model for Streams
When considering streams, the turbulent diffusion (i.e., longitudinal mixing) is generally insignificant

and equation (1) reduces to

$$\frac{dC}{dt} = -U\frac{dC}{dX} \pm \Sigma S \qquad (1a)$$

in which dc/dt is the change in concentration with time at a point source. Under steady-state conditions (i.e., there is no change in loading with time at a point source) equation (1a) becomes

$$U\frac{dC}{dX} = \pm \Sigma S = \frac{dC}{dt} \qquad (1b)$$

in which dc/dt is the change in oxygen concentration with time of flow downstream, U the velocity of flow, and X the distance downstream such that $t = x/U$. Assuming only deoxygenation by organic-matter oxidation and natural reaeration, equation (1b) becomes

$$\frac{dC}{dt} = U\frac{dC}{dX} - K_1 L + K_2(C_s - C) \qquad (1c)$$

Under steady-state conditions

$$\frac{dC}{dt} = 0 \quad \text{and} \quad 0 = U\frac{dC}{dX} - K_1 L + K_2(C_s - C) \qquad (1d)$$

Dividing throughout by velocity yields

$$\frac{dC}{dX} = \frac{K_1 L}{U} - K_2/U(C_s - C)$$

and integrating yields

$$C = C_s - \frac{j_1 L_0}{j_2 - j_r}(e^{-j_r X} - e^{-j_2 X}) - (C_s - C_0)e^{-j_2 X} \qquad (1e)$$

where $j_1 = K_1/U$, $j_2 = K_2/U$, and $j_r = K_r/U$. The critical point in the oxygen sag can be evaluated

$$K_2 D_c = K_1 L_c = K_1 L_0 \cdot 10^{-K_r t_c} \qquad (1f)$$

and

$$D_c = \frac{K_1}{K_2} L_0 \cdot 10^{-K_r t_c} \tag{1g}$$

by differentiating the oxygen-sag equation and equating to zero,

$$t_c = \frac{1}{K_2 - K_r} \log \frac{K_2}{K_r} \left[1 - \frac{D_0(K_2 - K_r)}{K_1 L_0} \right] \tag{1h}$$

The effect of the various sources and sinks on a hypothetical river are shown in Fig. 3.4.

Equation (1d) can be modified to include other sources and sinks of oxygen, as previously discussed. The coefficients that are employed in the oxygen-sag relationships are summarized in Table 3.4.

STREAM SURVEY AND DATA ANALYSIS

The stream survey should be conducted during a period when dissolved oxygen is available at all locations over the survey course. A low-flow high-temperature period is desirable. (This may not be possible in cases of a highly polluted stream.) Flood-flow conditions should be avoided.

Sampling stations should be selected for convenience of location but should encompass all pertinent stretches of the stream and must be below all new wastewater discharges, tributary confluences, and impoundments. It is important that complete mixing of all discharges with the streamwater be effected before sampling.

Physical Data to Be Collected
1. Stream cross section.
2. Stream flow (from nearest gaging station).
3. Sections of impoundment, marshes, rapids, etc.
4. Location, identification, and measurement of pollutional sources including waste flow variation and storm runoff.

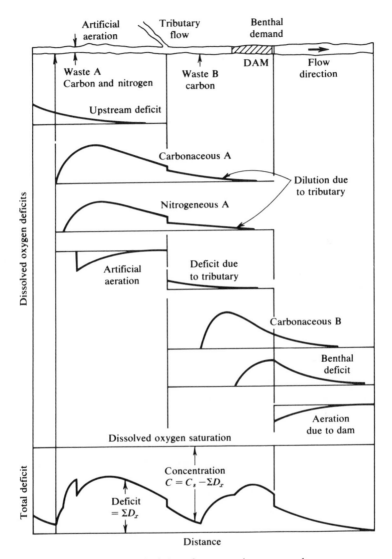

Figure 3.4 Sources and sinks of oxygen in a natural watercourse. (After O'Connor and Ditoro [1])

Sampling Data to Be Collected

1. Temperature.
2. Dissolved oxygen.
3. BOD (soluble and total).

**Table 3.4. Coefficients for the Evaluation of the
Assimilative Capacity of a Stream**

Coefficient	Definition	Dependent on	Temperature coefficient Θ
K_1	Oxidation-rate coefficient for soluble organics	Concentration and nature of organics remaining	1.065–1.075
K_r	Removal-rate coefficient for all organics; includes oxidation, sedimentation, and immediate demands	Concentration of total organics remaining	1.00–1.075
K_3	Removal rate by sedimentation	Concentration of settleable organics	—
K_2	Reaeration coefficient	Stream velocity and depth	1.028
K_n	Rate coefficient for nitrification	Nitrogen concentration present; concentration of organics and presence of secondary sewage effluent	1.106
k	BOD-bottle-rate coefficient	Nature of organics, i.e. raw waste or treated effluent	1.065–1.075

4. Suspended and volatile suspended solids.
5. Nitrogen (ammonia and nitrates).
6. Chlorophyl (in areas where algae are present).
7. Bottom deposits (where applicable).
8. Velocity of stream flow.

DEVELOPMENT OF THE COEFFICIENTS FROM THE SURVEY DATA

Conversion to Ultimate BOD The carbonaceous BOD data should be converted to ultimate values for evaluation of deoxygenation rates. Long-term BOD's should be run at pertinent stream stations (that is, below each major waste discharge) and the k and L_0 computed by one of the standard procedures (graphical, method of moments, etc.). In general, untreated wastes will have a k rate > 0.15 and treated wastewater a k rate < 0.1 (see Fig. 3.5).

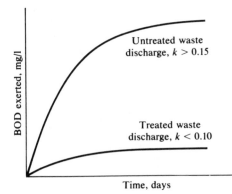

Figure 3.5 Variation in BOD exertion for treated and untreated wastes in a stream.

Deoxygenation Rate $—K_1$ The deoxygenation rate, K_1, is determined from the slope of a plot of the log BOD_u vs. time of passage or distance. (This plot can be developed either as lb of BOD_u or concentration of BOD_u after mixing in the stream flow.) K_1 is determined over a stream stretch where only soluble BOD is exerted (see Fig. 3.6). As shown in Fig. 3.6, K_1 should be readjusted below each waste outfall.

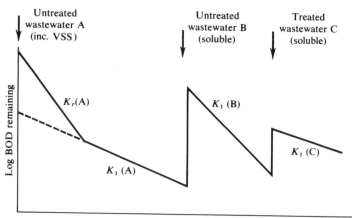

Figure 3.6 Deoxygenation relationships in a stream.

Total Removal Rate—K_r When volatile suspended solids are present, removal of BOD in the stream results both from oxidation and sedimentation of suspended organics on the stream bottom. The total removal rate is shown in Fig. 3.6. The coefficient, K_3, the sedimentation rate, is computed as $K_3 = (K_r L - K_1 L_1)/L_3$ in which L_1 and L_3 are the BOD in soluble and suspended form respectively. When no solids are present in the wastewater discharge, $K_r = K_1$.

Nitrification—K_N The calculations of K_N are shown in Fig. 3.7. The nitrification rate should be converted to terms of oxygen.

Upstream and Tributary Oxygen Contribution Oxygen will be supplied from that present in upstream water and in tributary flows. At each tributary point or location of stream aeration, the oxygen content in the stream should be recalculated after mixing with the main stream flow.

Reaeration The reaeration coefficient K_2 should be computed from equation (5) and corrected for stream temperature. The coefficients used in the equation should be selected from data developed on a similar stream.

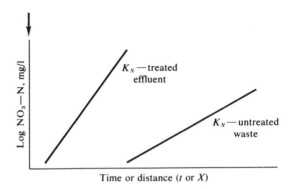

Figure 3.7 Nitrification rates in a stream.

Photosynthesis In areas where algae are present, photosynthesis will result in an increase in dissolved oxygen during the daylight hours and a decrease during the night hours. Where photosynthesis is known to exist, survey data should be collected on a diurnal cycle. Photosynthesis estimation can best be made from chlorophyll measurements. O'Connor and Di Toro[1] have shown that the photosynthesis effect can frequently be mathematically exposed as a sine fraction.

Benthal Demand Benthal demand is estimated from an undisturbed core sample of bottom deposit. Benthic demands are usually expressed as g of O_2/m^2 of stream bottom.

CORRELATION OF THE SURVEY DATA

The sources and sinks of oxygen are additive to develop the net oxygen balance in the stream [equation (1)]. These can be individually plotted and the net oxygen-sag curve computed by the addition and subtraction of the sources and sinks of oxygen as shown in Fig. 3.4. The computed curve should then be compared to the observed field-dissolved oxygen measurements. If the agreement is poor, the computed coefficients should be readjusted to close the difference between the observed and calculated results.

CALCULATION OF ASSIMILATIVE CAPACITY OF A STREAM

The stream survey is used to develop the coefficients defining the sources and sinks of oxygen. These data are then used to define (1) the oxygen-sag relationship under other conditions of stream flow and temperature, (2) the effect on the oxygen sag of additional sources of pollution, or (3) the degree of treatment necessary for existing or future sources of pollution to meet specified dissolved oxygen levels under defined conditions of temperature and stream flow. To meet these requirements, the physical factors involved, such as stream flow and depth and the reaeration-rate coefficients, must be readjusted to meet the revised conditions.

Variation in Physical Parameters   When the stream flow is changed to meet critical conditions, it is reasonable to assume that the velocity and depth will also be affected, which in turn influence the reaeration coefficient. General relationships have been developed for the hydraulic characteristics of a stream[10]:

$$H \sim Q^b$$
$$U \sim Q^f$$
$$W \sim Q^m$$

The sum of the exponents should be equal to 1.0. Correlations for a given river can be determined as shown in Fig. 3.8.

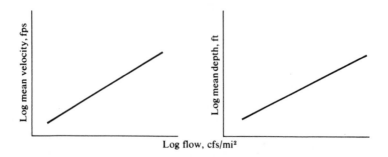

Figure 3.8 Hydraulic characteristics of natural streams.

Variation in Rate Coefficients The rate coefficients for sources and sinks of oxygen developed from the survey will be affected by changes in temperature, characteristics of waste discharges, and degree of waste treatment. The modifications of the coefficients needed for the prediction of other stream conditions are summarized in Table 3.5.

Table 3.5. Changes in Rate Coefficients as a Result of Environmental Effects

Coefficient	Temperature Coefficient	General Effects
K_R	1.0–1.075	Reduced by removal of suspended solids by treatment; temperature coefficient increases as $K_R \longrightarrow K_1$
K_1	1.075	For complete removal of suspended solids $K_R = K_1$; increases by addition of readily assimilable waste; decreases with degree of treatment of wastewater; increases with increasing stream fertilization
K_n	1.106	Increases with degree of treatment (domestic sewage); low for nitrified effluent
K_2	1.028	Decreases with the addition of untreated wastes or surface-active agents; increases with degree of treatment of wastewaters; increases with velocity; decreases with increasing depth

Calculation of Assimilative Capacity of a Stream

1. Estimate the low-flow-stream condition. This will usually be selected from a statistical analysis of drought flow conditions. Estimate the velocity and depth from a plot such as Fig. 3.8. (In the absence of data on the stream in question, H may be assumed to vary as $Q^{0.6}$ and V as $Q^{0.4}$.)

2. Revise the coefficient as follows:

K_R—correct for temperature and removal of suspended solids by treatment

K_1—correct for temperature and *estimated* degree of treatment

K_n—correct for temperature and *estimated* degree of treatment

K_2—correct for temperature, velocity, and depth at low-stream-flow condition; added wastes; or degree of treatment

Modify benthic demand and algae for projected conditions.

3. For the maximum allowable oxygen deficit, compute the maximum BOD loadings that can be discharged at each outfall from equation (1h) or modified for other sources and sinks of oxygen. The present and projected oxygen-sag curve and rate coefficients are schematically illustrated in Fig. 3.9. For simplicity, a stream with a single wastewater discharge is shown.

An assumed value of D_0/L_0 must be used (D_0 will be known). At t_c, the critical deficit, D_c, is computed as ($C_S - C_A$), in which C_A is the minimum allowable dissolved oxygen at the sag point. D_c can be computed in lb/day from the calculated flow.

At the critical point:

$$K_1 L_c = K_2 D_c$$

and

$$L_0 = L_c e^{K_r t_c}$$

The computed L_0 should be checked against the original assumption (and recalculated if necessary). From this value the required treatment can be computed.

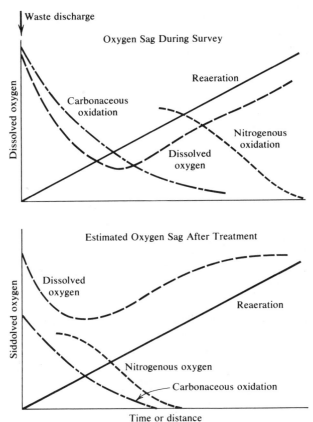

Figure 3.9 Oxygen sag before and after treatment.

DEVELOPMENT OF THE BOD–OXYGEN SAG MODEL FOR ESTUARIES

An estuary is defined as that portion of a river which is under the influence of tidal action in which the dispersion factor in equation (1) is always significant. The advective term may or may not be significant, depending upon the rate of freshwater flow and the cross-sectional area.

Estimation of the Dispersion Coefficient, ϵ For conservative substances such as chlorides or sulfates, equation (1) at a point source becomes

$$O = \epsilon \frac{\partial^2 C}{\partial X^2} - U \frac{\partial C}{\partial X} \qquad (12)$$

which integrates to

$$C = C_0 e^{Ux}/\epsilon \qquad (12a)$$

when $C = C_0$ at $x = 0$. Equation (12a) may be employed to estimate the turbulent diffusion coefficient ϵ from salinity or similar data. The velocity, U, is determined from the freshwater flow, Q, and the area, A, and the concentration, c, may be measured at various locations, x. The dispersion coefficient, ϵ, is calculated from the slope of a plot of the logarithm of the concentration vs. distance, X.

Estimation of the BOD Profile For a nonconservative substance with a first-order decay, under steady-state conditions, equation (1) becomes

$$O = \epsilon \frac{\partial^2 L}{dX^2} - U \frac{dL}{dX} - KL \qquad (13)$$

Equation (13) integrates to

$$L = L_0 e^{-\frac{Ux}{2\epsilon}(1 \pm m)} \qquad (13a)$$

in which

$$m = \sqrt{1 + \frac{4K\epsilon}{U^2}} \qquad (13b)$$

for the condition $L = L_0$ at $x = 0$ and $L = 0$ at $x = \pm\infty$. The concentration at $x = 0$ is obtained by taking a mass balance at this point. Equation (13a) may be expressed

$$L = L_0 e^{-j_1 X}$$

Estimation of the Dissolved Oxygen Profile The steady-state distribution of dissolved oxygen is defined from equation (1) as

$$O = \epsilon \frac{\partial^2 C}{\partial X^2} - U \frac{\partial C}{\partial X} + K_2(C_s - C) - K_1 L \tag{14}$$

Equation (14) integrates to

$$D = F \frac{W}{Q} \left(\frac{e^{j_1 x}}{m_1} - \frac{e^{j_2 x}}{m_2} \right) \tag{14a}$$

in which m_1 and m_2 were computed from equation (13b) and in which D = dissolved oxygen deficit; $F = K_1 / K_2 - K_r$,

$$j_1 = \frac{U}{2\epsilon} \left(1 - \sqrt{1 + \frac{4K_1\epsilon}{U^2}} \right)$$

$$j_2 = \frac{U}{2\epsilon} \left(1 - \sqrt{1 + \frac{4K_2\epsilon}{U^2}} \right)$$

Equation (14a) is identical in form to equation (1e) except that it contains the dispersion coefficient ϵ. In estuaries where the freshwater flow, U, is very small and can be neglected, equation (13) becomes

$$O = \epsilon \frac{\partial^2 L}{\partial X^2} - K_1 L \tag{15}$$

which integrates to

$$L = L_0 \exp(- X \sqrt{K_1/\epsilon}) \tag{15a}$$

The dissolved oxygen profile is computed by modifying equation (14):

$$O = \epsilon \frac{\partial^2 C}{\partial X^2} + K_2(C_s - C) - K_1 L \tag{16}$$

which integrates to

$$D = FL_0 \left(e^{\pm X\sqrt{K_1/\epsilon}} \right) - \sqrt{\frac{K_1}{K_2}} e^{\pm X\sqrt{K_2/\epsilon}} \qquad (16a)$$

L_0 is estimated as

$$L_0 = 2A \frac{W}{\sqrt{K_1 \epsilon}}$$

in which W is the mass of BOD discharged.

References

1. D. J. O'Connor and D. M. Di Toro, "An Analysis of the Dissolved Oxygen Variation in a Flowing Stream," *Advances in Water Quality Improvement,* University of Texas Press, Austin, Texas, 1967.
2. D. J. O'Connor and W. E. Dobbins, "Mechanism of Reaeration in Natural Streams," *Trans. ASCE,* **123,** 641 (1958).
3. W. E. Dobbins, "Nature of the Oxygen Transfer Coefficient in Aeration Systems," in *Biological Treatment of Sewage and Industrial Wastes,* Vol. 1, J. McCabe and W. W. Eckenfelder, eds., Reinhold, New York, 1956.
4. A. Gameson et al., "Some Factors Affecting the Aeration of Flowing Water," *Water Sanit. Eng.,* **6,** No. 2, 52 (1956).
5. P. A. Krenkel and G. T. Orlob, "Turbulent Diffusion and the Reaeration Coefficient," *J. Sanit. Eng. Div.,* **88** No. SA2, 53 (1962).
6. W. E. Dobbins, "Mechanism of Gas Absorption by Turbulent Liquids," *Advances in Water Pollution Research,* Vol. 1, Pergamon Press, Oxford, 1962.
7. W. E. Dobbins, "BOD and Oxygen Relationships in Streams," *J. Sanit. Eng. Div., ASCE,* **90,** No. SA3, 53 (1964).
8. M. A. Churchill et al., "Prediction of Stream Reaeration Rates," *J. Sanit. Eng. Div., ASCE,* **88,** No. SA4, (1962).
9. W. W. Eckenfelder and D. J. O'Connor, *Biological Waste Treatment,* Pergamon Press, Oxford, 1961.

10. R. J. Frankel, *Economic Evaluation of Water Quality— An Engineering-Economic Model for Water Quality Management,* First Annual Report, SERL Rep. 65-3, University of California, 1965.
11. K. Bosko, Discussion, *Advances in Water Pollution Research,* Water Pollution Control Federation, Washington D. C., 1967.
12. W. B. Langbein and W. H. Durum, "The Aeration Capacity of Streams," *U.S. Geol. Survey Circ.* No. 542, Washington, D. C. (1967).
13. A. L. Downing et al., "Nitrification in the Activated Sludge Process," *J. Proc. Inst. Sewage Purification,* 130 (1964).
14. Department of Scientific and Industrial Research, Water Pollution Research, 1963.

4

Characteristics of Municipal Sewage

Municipal sewage is composed of organic matter present as soluble, colloidal, and suspended solids. The pollutional contributories in sewage are usually expressed as a per capita contribution. A study of data reported by 73 cities in 27 states in the United States[1] during the period 1958–1964 showed a sewage flow of 135 gal/capita/day (511 l/capita/day) and a BOD_5 and suspended solids content of 0.2 lb/capita/day (90.7 g/capita/day) and 0.23 lb/capita/day (104 g/capita/day), respectively. The average composition of municipal sewage is shown in Table 4.1. It should be recognized that the presence of industrial wastes in the municipal system may radically alter

Table 4.1. Average Characteristics of Municipal Sewage[a]

Characteristics	Maximum	Mean	Minimum
pH	7.5	7.2	6.8
Settleable solids, mg/l	6.1	3.3	1.8
Total solids, mg/l	640	453	322
Volatile total solids, mg/l	388	217	118
Suspended solids, mg/l	258	145	83
Volatile suspended solids, mg/l	208	120	62
Chemical oxygen demand, mg/l	436	288	159
Biochemical oxygen demand, mg/l	276	147	75
Chlorides, mg/l	45	35	25

[a] After Hunter and Henkelekian [3].

58

these concentrations. In addition, these concentrations may be expected to vary by about a ratio of 3 over a 24-hr period. The chemical characteristics of sewage are summarized in Table 4.2. The general composition of municipal sewage has been re-

Table 4.2. Chemical Characteristics of Municipal Sewage[a]

Constituent	Type	Concentration
Volatile acids	Formic, acetic, propionic, butyric, and valeric	8.5–20 mg/l
Nonvolatile soluble acids	Glutaric, glycolic, lactic, citric, benzoic, and phenyllactic	Any acid 0.1–1.0 mg/l
Higher fatty acids	Palmitic, stearic, and oleic	$\frac{2}{3}$ fatty acid content
Proteins and amino acids	At least 20 types	45–50% of the total nitrogen
Carbohydrates	Glucose, sucruse, lactose, some galactose, and fructose	—

[a] After Walter [2].

ported by Hunter and Heukelekian[3] and is summarized in Table 4.3. Sewage volumes and BOD from various services are summarized in Table 4.4.

Table 4.3. Municipal Sewage Composition[a]

Fraction, %	Total Solids	Organic Matter[b]	Nitrogenous Matter[c]
Settleable	18	30	23
Supracolloidal	11	19	34
Colloidal	7	13	11
Soluble	64	38	22

Strength Parameter	Particulates, %	Solubles,%
Total solids	34.7	65.3
Volatile solids	57.6	42.4
COD	77.3	22.7
Organic nitrogen	80.5	19.5

[a] From Hunter and Henkelekian [3].
[b] Volatile solids or organic carbon.
[c] Organic nitrogen data.

Table 4.4. Sewage Volume and BOD for Various Services[a]

Type	Volume		5-day BOD	
	gal/capita/day	liter/capita/day	lb/capita/day	g/capita/day
Airports				
Each employee	15	56.8	0.05	22.7
Each passenger	5	18.9	0.02	9
Bars				
Each employee	15	56.8	0.05	22.7
Plus each customer	2	7.6	0.01	4.5
Camps and resorts				
Luxury resorts	100	378	0.17	77.2
Summer camps	50	189	0.15	68.1
Construction camps	50	189	0.15	68.1
Domestic sewage				
Luxury homes	100	378	0.20	90.7
Better subdivisions	90	340	0.20	90.7
Average subdivisions	80	302	0.17	77.2
Low-cost housing	70	265	0.17	77.2
Summer cottages, etc.	50	189	0.17	77.2
Apartment houses	75	284	0.17	77.2
(*Note:* If garbage grinders installed, multiply BOD factors by 1.5.)				
Factories (exclusive of industrial and cafeteria wastes)	15	56.8	0.05	22.7
Hospitals				
Patients plus staff	150–300 (avg 200)	578–1136 (avg 757)	0.30	136

[a]After Goodman and Foster [4].

Type	gal/capita/day	liters/capita/day
Hotels, motels, trailer courts, boarding houses (not including restaurants or bars)	50	189
Milk plant wastes	100–225 gal/1000 lb of milk	835–1880 l/1000 kg of milk
Offices	15	56.8
Restaurants		
Each employee	15	56.8
Plus each meal served	3 (per meal)	11.4 (per meal)
If garbage grinder provided, add	1 (per meal)	3.8 (per meal)

Table 4.4   (continued)

Schools	Elem.	High	Elem.	High
Day schools				
Each person, student or staff	15	20	56.8	76
Add per person if cafeteria has garbage grinder	—	—	—	—
Boarding schools		75		284
Swimming pools				
Employees, plus customers		10		38
Theatres				
Drive-in theatre per stall		5		19
Movie theatre per seat		5		19

Type	g 5-day BOD/capita/day		lb 5-day BOD/capita/day	
Hotels, motels, trailer courts, boarding houses (not including restaurants or bars)	68.1		0.15	
Milk plant wastes	560–1660/g/1000 kg of milk		0.56 to 1.66/1000 lb of milk	
Offices	22.7		0.05	
Restaurants				
Each employee	27.2		0.06	
Plus each meal served	13.6 (per meal)		0.03 (per meal)	
If garbage grinder provided, add	13.6 (per meal)		0.03 (per meal)	
Schools	Elem.	High	Elem.	High
Day schools				
Each person, student or staff	18	22.7	0.04	0.05
Add per person if cafeteria has garbage grinder	4.5	4.5	0.01	0.01
Boarding schools	77.2		0.17	
Swimming pools				
Employees plus customers	13.6		0.03	
Theatres				
Drive-in theatre per stall	9		0.02	
Movie theatre per seat	9		0.02	

References

1. R. C. Loehr, "Variation of Wastewater Parameters," *Public Works*, **99**, 81 (1968).
2. Leo Walter, "Composition of Sewage and Sewage Effluents," Parts 1 and 2, *Water Sewage Works,* **108,** Nos. 11 and 12, 428–478 (1961).
3. J. V. Hunter and H. Henkelekian, "The Composition of Domestic Sewage Fractions," *J. Water Pollution Control Federation,* **37,** No. 8, 1142 (1965).
4. B. Goodman and J. W. Foster *Notes on Activated Sludge*, Smith and Loveless Co., Lenexa, Kans., 2nd ed., 1969.

5

Industrial Wastes

Industrial wastes represent a major pollutional problem in industrial societies. The technology of treatment and control of these wastes has been largely developed over the past decade. The procedure for generating data needed for the design of waste-treatment facilities is developed in this chapter. A discussion of the wastewater characteristics, water-reuse and waste-recovery practices, and waste treatment for the major wastewater-producing industries is included. Many of the data presented are summarized from the series of volumes, *The Cost of Clean Water*, published by the Department of the Interior in 1968. The reader is referred to these references for more detailed study of specific industries.

PROCEDURE FOR INDUSTRIAL-WASTE SURVEYS

The purpose of the industrial waste survey is to establish present waste loadings and flows, pinpointing major individual sources of pollution within the industrial establishment; locate and define water-elimination and reuse possibilities and product-recovery and waste-reduction sources; and establish a material balance and flow diagram of all major pollutants before and after volume and strength reduction. The procedure to be followed is listed below:

1. Review process-water use and sources of waste. All primary sources of pollution should be established. Type and duration of operation should be defined, for example, batch processes discharging several times a day, continuous process operation, and intermittant operation of a process.

63

2. Define raw-materials makeup and wastewater characteristics for all processes delineated in step 1.

3. Develop a sewer map of the plant.

4. Locate sampling and measurement stations; these should include all significant sources of waste (see step 1). Sampling stations should be located where the wastewater flow can be measured or estimated.

5. Install total effluent flow-measurement device.

6. Select analyses to be run (this will be indicated by step 2). Set sampling and analyses schedules. Batch processes should be composited over the duration of each discharge and weighted according to flow; continuous processes should be composited and weighted according to flow. The compositing schedule will depend upon the variability in waste characteristic with respect to proposed treatment facility operation or effect on the receiving water (see Table 5.1).

Table 5.1. Suggested Sampling or Compositing Schedule

Characteristic	High Variability	Low Variability
BOD[a]	4 hr	12 hr
COD or TOC[a]	2 hr	8 hr
Suspended solids	8 hr	24 hr
Alkalinity or acidity	1-hr grab	8-hr grab
pH	Continuous	4-hr grab
Nitrogen and phosphorus[b]	24 hr	24 hr
Heavy metals	4 hr	24 hr
Temperature	2 hr	8 hr

[a] The compositing schedule where continuous samplers are not used depends on variability, i.e., 15 min for high variability to 1 hr for low variability.

[b] Does not apply to nitrogen or phosphorus wastes (e.g., fertilizer).

7. Define the duration of the sampling program. For wastes of relatively uniform composition (e.g., those from the paperboard industry) several days should be sufficient to establish design criteria. For plants with a variety of products and production schedules, such as a diverse chemical plant,

the survey should be programmed to cover all major production schedules.

8. Develop a flow and material balance for the plant from the data developed in steps 1 through 7.

9. Develop statistical plots for all significant characteristics for both the total plant flow and all major individual process contributions. Where possible, these statistical plots should be related to production, that is, gal/ton of product or lb of BOD/ton of product. This permits extrapolation to other production schedules.

10. Locate sources for waste segregation, water reuse, product recovery, recirculation, etc. This should be done using the data from steps 1, 2, 8, and 9.

11. Reestimate a flow and material-balance diagram (step 8) and probable statistical variation (where possible).

12. Establish possibilities for waste segregation, separate treatment, etc. This will usually be based on flow, concentration, and composition. Examples are segregation of noncontaminated cooling waters and segregation and separate treatment of high-concentration toxic wastes.

Flow measurement during the industrial-waste survey can be achieved by a number of methods, depending on the nature and accessibility of the samplings point. These are

1. Installation of a weir—this can be done in gutters, channels, and partially filled sewers at a manhole.

2. Bucket and stopwatch—this is applicable to low flow rates.

3. Pumping duration and rate—must be estimated from the characteristic curves of a pump.

4. Timing a floating object between two manholes—applicable to partially filled sewers; depth of flow in the sewer must also be measured. The average velocity is equal to 0.8 times the surface velocity. Flow rate is computed from the continuity equation, $Q = AV$.

5. Plant water-use records—account must be taken of water losses in product and due to evaporation.

6. Timing change of level in tank or reactor—used primarily for batch-operation discharges.

CORRELATION OF INDUSTRIAL-WASTE-SURVEY DATA

Most industrial wastewater discharges are highly variable in volume and composition and are susceptible to statistical analysis. These frequency-concentration relationships are necessary both to determine water-reuse and process modifications and to define the criteria for waste-treatment process design.

Statistical Correlation For small amounts of data (i.e., less than 20 datum points), the procedure is as follows:

1. Arrange the data in increasing order of magnitute [col. 1 of Table 5.2].

Table 5.2.

(1) BOD mg/1	(2) m	(3) Plotting Position
200	1	5.55
225	2	16.65
260	3	27.75
315	4	38.85
350	5	49.95
365	6	61.05
430	7	72.15
460	8	83.75
490	9	94.35

2. m is the total number of values and n the assigned serial number from 1 to n.

3. The plotting position is determined by dividing the total number of samples into 100 and assigning the first value as one half this number [col. (3)].

These data are illustrated in Fig. 5.1.

When large numbers of data are to be analyzed it is convenient to group the data for plotting, e.g., 0 to 50, 51 to 100, 101 to 150, etc. The plotting position is determined as $m/n + 1$, where m is the cumulative number of points and n the total number of observations. The statistical distribution of data serves several important functions in developing the industrial-water-management program.

1. Fluctuations in waste discharge from unit process operations may indicate procedures for dampening or reducing the pollutional losses.

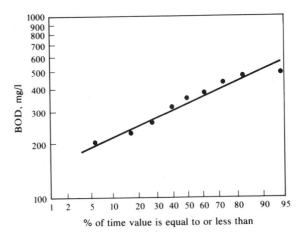

Figure 5.1 Statistical variation in BOD of a wastewater.

2. Variation of waste discharge from the total plant operations will indicate the need for equalization prior to treatment.

3. Depending on the treatment process employed, various frequency values are used to design. For example, the hydraulic capacity of the plant should be in excess of the 99 percent frequency, whereas sludge-handling facilities are usually designed on the average value (50 percent).

Waste flow and loadings are usually related to a suitable unit of production, e.g., gal/ton or lb/ton. Values for particular industries are reported in the appropriate section. This permits comparison of pollutional loads with industries having comparable processes, and also extrapolation to increased or expanded production within the plant. Pollutional load will also be related to the level of technology of the industry, generally decreasing as the level of technology increases. These comparisons are shown on pages 88 and 96 for petroleum and pulp and paper mill wastes, respectively.

Waste-Reduction Practices Frequently, relatively simple in-plant modifications can result in substantial reduction in volume and strength of wastes to be treated and hence reduction in plant construction and operating costs. They are summarized below.

1. Recirculation. Relatively noncontaminated waters can be recirculated with no treatment or minimial treatment; examples are produce washing in canneries and gutter or channel water for transporting product or wastes.

2. Segregation. Cooling water and other noncontaminated flows can be segregated prior to treatment; high-strength or toxic wastes can be segregated for separate treatment.

3. Removal in semidry state. Concentrated residues can be removed semidry for disposal, eliminating flushing to the sewer. This has been used in canneries for the removal of cooking residues from kettles and the removal of residues from beer storage vats in breweries.

4. Reduction. Use of automatic cutoffs on hoses and use of spray rinses; these practices have substantially reduced wastewater volumes in dairies and plating plants.

5. Substitution. The substitution of chemical additives of a lower pollutional effect in processing operations, e.g., substitution of surfactants for soaps in the textile industry.

An economic balance for water reuse must consider the cost of raw water, the cost of wastewater treatment to suitable process quality requirements, and the cost of wastewater treatment to a receiving water. This balance is illustrated in the following example.

Example: A plant uses 10,000 gal/hr of process water with a maximum contaminant concentration of 1 lb/1000 gal. The raw water supply has a contaminant concentration of 0.5 lb/1000 gal (see Fig. 5.2). Optimize a water-reuse system for this plant based on a raw water cost of 20 cents/1000 gal. The following conditions apply:

Evaporation and product loss (E)	1000 gal/hr
Contaminant addition (Y)	100 lb/hr
Maximum discharge to receiving water	20 lb/hr

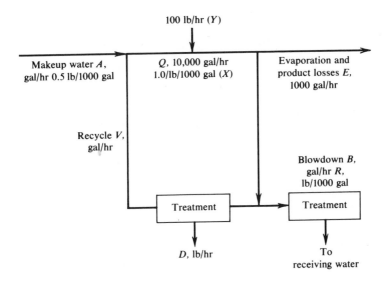

Figure 5.2 Water reuse and treatment balance.

Example calculation: Let $A = 3000$ gal/hr. Since $A = E + B$, $B = 3000 - 1000 = 2000$ gal/hr and the recycle, V, is 7000 gal/hr. By a material balance

$$\frac{(A + V)X + Y}{(A + V) - L} = R$$

$$\frac{(10)(1.0) + 100}{10 - 1} = \frac{110}{9}$$

The removal for reuse can be computed as

$$100 + (3)(0.5) - (2)(12) = D$$

The required efficiency of treatment is

$$\frac{77.5}{(7)(12)} \times 100 = 91\%$$

The total blowdown is $B - R$ or $(2)(12) = 24$ lb/hr. The required treatment efficiency for discharge to the river is

$$\frac{24 - 20}{24} \times 100 = 16.7\%$$

The raw water cost for 3000 gal/hr is $(3)(24)(0.20) = \$14.40/$day. The cost of effluent treatment for discharge to the river is 5 cents/1000 gal (see Fig. 5.3) for a total daily cost of $(2)(0.05)$

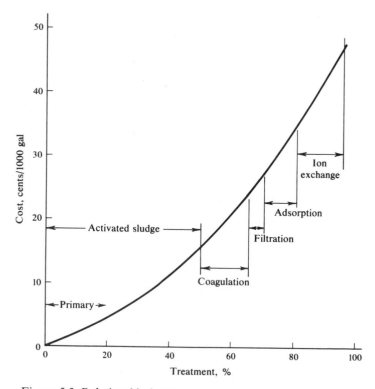

Figure 5.3 Relationship between total water cost and treatment.

(24), or $2.40/day. The cost of treatment for reuse (see Fig. 5.3) is $(7)(0.42)(24) = \$70.50/$day. The net total cost is $87.30/day. A similar series of calculations is made for freshwater inputs varying from 2000 to 10,000 gal/hr (no reuse).

The total daily water cost can then be plotted versus percent recycle as shown in Fig. 5.4.

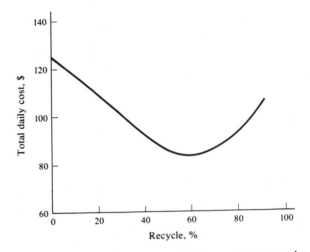

Figure 5.4 Relationship between total daily water cost and treated waste recycle for reuse.

BREWERY WASTES

General Brewing consists of several steps, resulting in the production of finished beer (see Fig. 5.5).[1]

Brewing Process Malt is sent to the mash tun, where it is mixed at elevated temperatures with a measured amount of water. The malt is first mashed, and then the water, at 150 to 160°F, is added. The malt undergoes enzymatic changes, following which the mash is transferred to the lauter tun, where settling and screening takes place. The liquor is decanted off into the brewing kettle or "copper," where, after hops have been added, it is boiled. The spent grains from the lauter tun are next removed to a hopper, and in one case amounted to 12.2 lb (5.52 kg dry basis) per barrel of beer produced. The liquor from the brewing kettle, or wort, is sent to the hopjack filter, where the spent hops are removed. The hops amount to about 0.5 lb (0.22 kg dry basis) per barrel of beer produced.

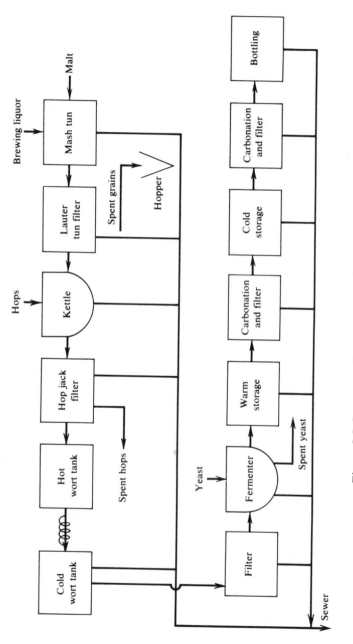

Figure 5.5 Sequence of process operations in a brewery.

The wort is then sent to a holding tank, called the "hot wort" tank. The liquor is cooled and sent to the "cold wort" tank for storage. The wort from storage is filtered through a diatomaceous-earth filter from which it passes to the fermentation tank, which is maintained at 61 to 70° F. Yeast is added and fermentation takes place. Surplus yeast and proteinaceous substances denatured from the fermentation solutions have amounted to 0.53 lb (0.24 kg dry basis) per barrel of beer produced. The brew is then transferred to warm storage and then filtered through diatomaceous earth. The beer filter cake (diaomaceous earth, proteins, yeast cells, and hop resins) can amount to 1.0 lb (0.454 kg dry basis) per barrel of beer produced. The beer is sent from warm storage to carbonation and cold storage. It is again carbonated and filtered and finally packaged in bottles or kegs.

Waste and Wastewater Sources   The major part of the BOD and suspended solids emanate from the brewing operations, including fermentation, while a significant part of the volume is discharged from the bottling, pastuerizing, and kegging operations. The approximate distribution of wastewater from the various brewing steps is shown in Fig. 5.6. Average waste and wastewater quantities from several breweries are summarized in Table 5.3. The statistical variation in wastewater characteristics based on 25 reported sets of data are summarized in Table 5.4.[2] It should be noted that brewery wastes have a BOD/N/P ratio of 100:4:1, which renders them somewhat deficient in nitrogen for most biological treatment processes.

Water-Reuse and Waste-Recovery Practices   Water reuse and reduction of effluent volume has been achieved by
1. Use of three washes for kettles, and so on, with the third reuse for the subsequent first wash.
2. Reuse of cooling water in bottling and pastuerizing.
3. Countercurrent flow in pastuerizing and bottling.
Product recovery and waste reduction has been accomplished by
1. Mash and filter cloth washings from the filtration of spent mash and grains.

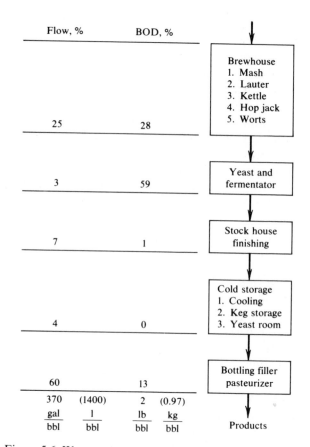

Figure 5.6 Wastewater volume in BOD sources in a brewery.

2. Drying waste liquid from spent grains. This high organic waste has been evaporated and pellitized and sold as an animal feed supplement. This amounts to about 0.8 ft³ (22.6 l) of spent grain/bbl and about 5.1 gal/bbl (19.3 l/bbl).

3. Filter and dry yeast from the fermenter. This amounts to about 50 bbl of settled yeast/1500 bbl of beer and is high in BOD and suspended solids.

4. Remove the sediment from the chill storage tanks as a slurry for separate disposal instead of flushing to the sewer.

5. Recovery of the diatomaceous earth from the filtration step.

Table 5.3. Brewery-Waste Characteristics

Brewery	Flow gal/bbl(l/bbl)	BOD lb/mg(kg/m³)	BOD lb/bbl (kg/bbl)	BOD mg/l	Suspended Solids lb/mg (kg/m³)	Suspended Solids lbs/bbl (kg/bbl)	Suspended Solids mg/l
A[a]	295[b](1110)	15,000 (1.8)	2.1 (0.95)	1832[c]	8550(1.03)	1.3 (0.59)	1028[d]
New York City study	—	—	1.6 (0.73)	—	—	0.6 (0.27)	—
B	170 (644)		1.5 (0.68)	1040		1.8 (0.82)	
C	350 (1320)		2.4 (1.1)	850		1.1 (0.50)	
Chicago breweries	320 (1210)		3.1 (1.4)	1160		1.7 (0.77)	
C[a]	130 (493)		1.7 (0.77)	1500		1.7 (0.77)	470

[a] Separates spent grains and spent hops.

[b] Brewhouse, 130 gal/bbl (493 l/bbl); bottling, 165 gal/bbl (625 l/bbl).

[c] Brewhouse, 3580 mg/l; bottling, 384 mg/l.

[d] Brewhouse, 2220 mg/l; bottling, 80 mg/l; spent hops removed.

Table 5.4.   Statistical Variation in
Brewery-Waste Characteristics[a]

	Mean	16%	84%
Flow			
gal/bbl	370	225	520
l/bbl	1440	850	1970
BOD			
lb/bbl	1.90	0.84	4.2
kg/bbl	0.86	0.38	1.9
Suspended solids			
lb/bbl	1.03	0.43	2.5
kg/bbl	0.47	0.195	1.13

[a] Based on a barrel of beer which contains 36.5 gal (117 l).

Brewery-Waste Treatment   Since many breweries are located in large metropolitan-areas, their wastes are treated in municipal systems. This usually involves a surcharge for BOD and suspended solids. Separate biological treatment of brewery wastes has been accomplished by trickling filtration, activated sludge, and, in a limited number of cases, anaerobic treatment.

BOD removals in excess of 90 percent have been achieved in trickling filters at loadings in the order of 75 gal/yd³/day (151 l/m³/day)[3]. This process is susceptible, however, to rapid changes in pH and shock organic loadings.

The activated sludge process has been successfully employed at a number of plants, resulting in BOD removals in excess of 90 percent. The completely mixed process is preferred to minimize the effects of shock loadings and to maintain a maximum food to microorganism ratio in all parts of the process [0.45 lb (kg) of BOD/day/lb (kg) of MLSS]. Excessive organic loadings will result in filamentous growths with poor settling properties. Nitrogen must usually be added as a nutrient supplement.

References
1. W. W. Eckenfelder and C. G. Bueltman, "Reducing Brewery Wastes BOD by Bio-oxidation Treatment," *Wastes Eng.*, **31** No. 1, 16 (1960).

2. J. T. O'Rourke and H. E. Tomlinson, "Effects of Brewery Wastes on Treatment," *Ind. Water Wastes,* **7,** No. 5, 119 (1962).
3. Ruben Schneider, "Waste Disposal at a Modern Brewery," *Sewage Ind. Wastes,* **22,** No. 10, 1307 (1950).

CANNERY WASTES

General   A differentiation will be made in this discussion between wastes resulting from canned fruits and vegetables and frozen fruits and vegetables.

The major steps involved in the processing of fruits and vegetables are initial preparation, which involves precleaning, size grading, and sorting; and converted product handling, which involves blanching, mixing additives, pulping, cooking, etc.

Wastes from initial preparation include soil, sand, stones, insecticides, vegetation, dried juices, and other residues which are washed from the product. The quantity of impurities will depend on the type of harvesting, increasing with mechanical crop picking. Initial preparation will include trimming, coring, pitting, and cutting, which is a mechanical process, resulting in a solid residue. Some liquid waste, such as juices and equipment wash water, results from cutting. Peeling by hand, steam, machines, or chemicals yields a large volume of wastewater with a high BOD. Peeling in a lye solution is common, resulting in a wastewater highly alkaline but with a low concentration of dissolved organic matter.

The final step in initial preparation transports the product by conveyor or flume to the converted product-handling process. Flume water is high in volume and organic load and is frequently reused. Initial preparation contributes up to 50 percent of the total wastewater flow, almost all the suspended solids, and a significant part of the organic load.

Converted product handling involves blanching, which exposes the product to steam or hot water to retard the enzymatic degradation of the organic matter. The wastewater from blanching is hot and contains dissolved organic matter.

Table 5.5. Daily Waste Quantities from Unit Processes[a]

| Process | Wasteload, | | (kg/day/case) | Wastewater Volume |
| | lb/day/case | | | (gal/case)(l/case) |
	BOD	SS	TDS	
Washing	0.050–0.3	0.05–0.4	300–1.500	08.0–25.0
	(0.023–0.140)	(0.023–0.18)	(140–680)	(30.3–94.7)
Belt conveyor	0.003–0.001	0.01–0.02	30–100	01.0–05.0
	(0.0014–0.00045)	(0.0045–0.0091)	(14–45)	(3.8–18.9)
Sorting, pitting, slicing, and so on	0.005–0.06	0.015–0.070	100–500	01.0–07.5
	(0.0045–0.027)	(0.0068–0.032)	(45–230)	(3.8–28.4)
Blanching and/or peeling	0.1–0.4	0.10–0.4	2,000–4,000	06.0–25.0
	(0.045–0.18)	(0.045–0.18)	(900–1800)	(22.7–94.7)
Exhausting of cans	0–0.015	0–0.015	0–200	0–02.5
	(0–0.0068)	(0–0.0068)	(0–90)	(0–9.5)
Processing	0.005–0.06	0.005–0.0450	40–1,200	01.0–05.0
	(0.0045–0.027)	(0.0045–0.020)	(18–540)	(3.8–18.9)
Cooling of cans	0.005–0.06	0.005–0.0300	300–1,000	06.0–30.0
	(0.0045–0.027)	(0.0045–0.014)	(140–450)	(22.7–113)
Plant cleanup	0.032–0.12	0.030–0.15	230–1,000	06.0–20.0
	(0.0145–0.054)	(0.014–0.068)	(104–450)	(22.7–75.8)
Box washing	0.01–0.025	0.015–0.04	200–500	02.0–05.0
	(0.0045–0.011)	(0.0068–0.018)	(91–230)	(7.6–18.9)
Total	0.26–1.05	0.23–1.17	3,200–10 000	31.0–125.0
	(0.12–0.48)	(0.10–0.53)	(1,400–4,540)	(113–475)
Average	0.7	0.8	7,500	075.0
	(0.32)	(0.36)	(3,400)	(284)

[a] From *The Cost of Clean Water* [1].

Pulping, straining, and cooking in vats may contribute some wastes from spillage. Considerable waste may result from the washing and cleansing of vats, pulps, etc.

A large volume of relatively clean water results from the canning and cooling operations. This water can be reused in flume transportation, product washing, and can cooling.

Waste and Wastewater Sources Wastewater volumes and characteristics from cannery operations will vary markedly depending upon the type of raw product, the extent of hydraulic transport of the product, recirculation of process waters, and in-plant transport of solid waste materials. Significant pollutants will result from hot lye peeling, sorting, slicing, etc.; washing and cooling of cans; and plant cleanup. The range of waste quantities from these operations is summarized in Table 5.5. The distribution of water use in cannery operation is summarized in Table 5.6. Typical canning wastes characteristics are summarized in Table 5.7.

The average waste characteristics from frozen-food preparation are 0.62, 0.63, and 0.50 lb (0.28, 0.284, and 0.227 kg) of BOD, SS, and TDS per case, respectively, with a daily wastewater volume of 125 gal (474 l)[1].

Table 5.6. Water Reuse[a]

Operation	Gross Water Use, %	
	Range	Average
Raw production preparation (including peeling)	25–55	34
Syrup and brine	2–10	6
Steam and sterilization	10–20	14
Cooling	6–60	35
Cleanup	5–25	8
Other	1–6	3

[a] From *The Cost of Clean Water* [1].

Water-Reuse and Waste-Reduction Practices Water reuse can be practiced in canning plants by recycling washwaters (the final rinsing can serve as makeup for the initial

Table 5.7. Typical Canning Wastes

Product	Waste Volume, gal/(l/case)	5-Day BOD		Suspended Solids	
		mg/l	lb/(kg/case)	mg/l	lb/(kg/case)
Apples	29–46 (110–174)	1680–5530	0.64–1.31 (0.29–0.60)	300–600	0.10–0.20 (0.045–0.091)
Apricots	65–91 (246–344)	200–1020	0.15–0.56 (0.068–0.25)	200–400	0.14–0.25 (0.064–0.11)
Cherries	14–46 (53–174)	700–2100	0.16–0.50 (0.073–0.23)	200–600	0.05–0.14 (0.023–0.064)
Cranberries	11–23 (42–87)	500–2250	0.10–0.21 (0.045–0.095)	100–250	0.02–0.05 (0.091–0.023)
Peaches	51–69 (193–261)	1200–2800	0.69–1.20 (0.31–0.55)	450–750	0.24–0.34 (0.11–0.15)
Pineapples	74 (280)	26	0.002 (0.00091)	—	—
Asparagus	80 (303)	16–100	0.01–0.07 (0.0045–0.032)	30–180	0.02–0.12 (0.091–0.055)
Beans, baked	40 (151)	925–1440	0.31–0.48 (0.14–0.22)	225	0.07 (0.032)
Beans, green wax	30–51 (114–193)	160–600	0.15–0.67 (0.068–0.30)	60–150	0.02 0.04 (0.091–0.018)
Beans, kidney	20–23 (76–87)	1030–2500	0.19–0.45 (0.086–0.20)	140	0.02 (0.091)
Beans, lima, dried	20–23 (76–125)	1740–2880	0.30–0.60 (0.14–0.27)	160–600	0.05–0.10 (0.023–0.045)

Beans, lima, fresh	57–294 (216–1110)	190–450	0.21–0.47 (0.095–0.21)	420	0.20–1.02 (0.091–0.46)
Beets	31–80 (117–303)	1580–7600	1.00–2.00 (0.45–0.91)	740–2220	0.50–1.00 (0.23–0.45)
Carrots	36 (136)	520–3030	0.11–0.67 (0.05–0.30)	1830	0.40 (0.18)
Corn, cream style	28–33 (106–125)	620–2900	0.17–0.66 (0.077–0.30)	300–675	0.07–0.17 (0.032–0.077)
Corn, whole kernel	29–80 (110–300)	1120–6300	0.74–1.50 (0.33–0.68)	300–4000	0.20–0.95 (0.091–0.43)
Mushrooms	—	76–850	4.77–53.38 (2.16–24.2)	50–240	3.14–15.07 (1.42–6.83)
Peas	16–86 (61–326)	380–4700	0.27–0.63 (0.12–0.29)	270–400	0.06–0.20 (0.27–0.091)
Potatoes, sweet	90 (340)	1500–5600	1.10–4.40 (0.50–2.0)	400–2500	0.31–1.95 (0.14–0.88)
Potatoes, white	—	200–2900	—	990–1180	—
Pumpkin	23–57 (87–216)	1500–6880	0.72–1.31 (0.33–0.59)	785–1960	0.38 (0.17)
Sauerkraut	3–20 (11–76)	1400–6300	0.10–0.30 (0.045–0.14)	60–630	0.01–0.10 (0.0045–0.045)
Spinach	180 (68)	280–730	0.42–1.11 (0.19–0.50)	90–580	0.14–0.88 (0.063–0.40)
Squash	23 (87)	4000–11,000	0.76–2.09 (0.34–0.95)	3000	0.57 (0.26)
Tomatoes	3–114 (11–432)	180–4000	0.11–0.17 (0.050–0.077)	140–2000	0.06–0.13 (0.027–0.059)

wash); flume water can be screened and recycled. Appropriate bacteriacides can be used to control bacterial growth. Spray rinses can be used as flume water. Cooling-tower recirculation, with blowdown used in various unit operations, will significantly reduce water usage.

Specific water-conservation measures may include

1. Automatic shutoff valves on water hoses.
2. High-pressure sprays.
3. Eliminate excessive overflow from soaking and washing tanks.
4. Substitute mechanical conveyors for water flumes.
5. Cooling-water recirculation.

Waste strength can be reduced by removal of dry material (pulp, peelings, etc.) by conveyor rather than by flume.

Waste-Treatment Practices   At the present time over 60 percent of cannery wastes are discharged to municipal systems. Since cannery wastes are readily biodegradable, they are compatible with biological treatment processes, although in some cases supplementary nutrients (nitrogen and phosphorus) must be added. The major problems encountered are the seasonal nature of many canneries, which can create overloads at the sewage-treatment plant, and slug discharges, which can lead to process upset. To minimize these problems, water reuse and proper pretreatment (screens and some equalization of concentrated wastes) should be instituted.

The most common methods of cannery-waste treatment in nonurban areas is lagooning and spray irrigation. Where anaerobic lagoons are employed, sodium nitrate may be required for odor control. Sodium nitrate addition has varied from 20 to 200 lb/1000 cases (9 to 90 kg/1000 cases) of No. 2 cans.

Spray irrigation is feasible where land area is available and is primarily limited by the capacity of the spray field to absorb the wastewater. Application rates of 40 to 250 gpm/acre (0.0339 to 0.295 l/min/m²) have been employed.

Biological treatment is feasible for cannery wastes but is frequently unattractive because of the seasonal nature of canning operations. Aerated lagoons have received the widest application. The efficiencies that might be expected from various treatment processes are summarized in Table 5.8.

Table 5.8. Treatment Removal Efficiency[a]

| Method | Pollution Reduction, % | | |
	Flow to Surface Water	BOD to Surface Water	SS to Surface Water
Screening, 20–40 mesh	0	0–10	56–80
Sedimentation	0	10–30	50–80
Flotation	0	10–30	50–80
Chemical precipitation	0	39–89	70–90
Chemical oxidation	—	—	—
Activated sludge	0	59–97	90–95
Trickling filtration	0	36–99	85–90
Anaerobic fermentation	0	40–95	—
Lagooning	0–50	83–99	50–99
Spray irrigation	50–100	100	100
Sand filtration	50–100	15–85	100

[a] From *The Cost of Clean Water* [1].

Reference

1. *The Cost of Clean Water*, Vol. III, Industrial Waste Profile No. 6, U.S. Department of Interior, FWPCA, Washington, D.C., 1968.

DAIRY WASTES

Wastewaters in the dairy industry originate primarily from tank and equipment washing, product spillage and losses. Water conservation, collection of drippings, and other practices can materially reduce the BOD discharged to the sewer. The wastewater characteristics for various dairy products are shown in Table 5.9.

Since many dairies are located in or near urban areas, treatment in municipal systems is common. Dairy wastes have also been treated by activated sludge and aerated lagoons and by spray irrigation.

Table 5.9. Dairy Wastewater[a]

Product	BOD lb/100 lb	Volume	
		gal/100 lb	l/100 kg
Creamery butter	0.34–1.68	410–1350	3410–11,300
Cheese	0.45–3.0	1290–2310	10,780–19,300
Condensed and evaporated milk	0.37–0.62	310–420	2590–3500
Ice cream[b]	0.15–0.73	620–1200	5180–1000
Milk	0.05–0.26	200–500	1670–4180

[a] From *The Cost of Clean Water* [1].
[b] Per 100 gal of product.

MEAT-PACKING WASTES

The meat-packing industry includes the slaughtering and processing of beef and hogs and may include the cooking, curing, smoking, and pickling of meat; the manufacture of sausage; the rendering of edible fats into lard and edible tallow; and the rendering of inedible fats into greases. The poultry-processing industry will be considered separately.

Old technology in the meat-packing industry would include blood recovery but wet cleanup with washing of paunch manure to the sewer and no evaporation of rendering tank water. Typical technology includes blood recovery, wet dumping of paunch manure with hauling of gross paunch material, dry rendering or evaporation of tank water, and wet cleanup. Advanced technology includes blood recovery, dry dumping of paunch manure, continuous dry rendering, and dry cleanup followed by wet cleanup.

The waste loading, classified by the technology, is shown in Table 5.10.

Most waste-treatment processes include catch basins with detention periods of 20 to 40 min, equipped with grease skimming as a pretreatment. BOD removal may average 25 percent. Air flotation with coagulants has been used for the removal of greases, fat, and some proteins, resulting in 50 percent BOD removal. Biological treatment has been most popular and effective. Anaerobic lagoons will yield 80 percent BOD

Table 5.10. Wastewater Characteristics per Thousand Pounds (Thousand Kilograms) of Liveweight Killed[a]

Technology	lb of BOD/1000 lb (kg of BOD/1000 kg)	gal/1000 lb	l/1000 kg
Old	20.2	2112	17,600
Typical	14.4	1294	10,800
Advanced	11.3	1116	9,200

[a] Post catch basin.

reduction, with the aerobic processes providing an excess of 90 percent removal. (An anaerobic lagoon followed by an aerobic lagoon will effect 95 percent reduction.)

Poultry Processing Wastes from poultry processing originate primarily from blood, feathers, and cleanup. After arriving at the processing plant, the birds are slaughtered, followed by scalding and defeathering. The carcasses are then eviscerated and packed. The older plants will not recover blood and use water, feathers, and offal from the processing area. The newer plants will recover blood and other solid wastes. The waste loads are summarized in Table 5.11.

Table 5.11. Wastewater Characteristics per 1000 Birds

Plants	lb of BOD/1000 birds	kg of BOD/1000 birds	gal of BOD/1000 birds	l of BOD/1000 birds
Older	31.7	14.4	4000	15,100
Newer	26.0	11.7	7300	27,600

The majority of poultry-processing plants use municipal treatment facilities with screening as a pretreatment for the removal of feathers and offal. Other treatment considerations would be similar to that employed in the meat-packing industry.

PETROLEUM-REFINERY WASTES

General A petroleum refinery is a complex combination of interdependent processes. The principal operations, from the point of view of water-pollution control, are

1. Crude-oil and product storage. During crude-oil storage water separates from the oil, resulting in a wastewater containing free and emulsified oil, suspended solids, and a bottom sludge. Wastewaters from intermediate storage contain polysulfides and iron sulfides. Finished product storage can produce alkaline wastewaters high in BOD and tetraethyllead. Tank cleaning can yield large quantities of oil, COD, and suspended solids.

2. Crude desalting by chemical or electrostatic means is employed for the removal of inorganic salts, suspended solids, arsenic, and other impurities, which may poison catalytic cracking catalysts. Wastewaters are high in COD and BOD and contain emulsified and free oil, ammonia, phenol, and suspended solids as well as a high chloride content.

3. Crude-oil fractionation is a primary refining process to separate the crude petroleum into intermediate fractions of various boiling-point ranges, including gasoline, naptha, kerosene, fuel oil, and asphalt. Wastewaters include the water drawn off from overhead accumulators prior to recirculation or transfer of the hydrocarbons. These wastes are high in sulfides, oil, chlorides, mercaptans, and phenols. Wastewaters containing emulsions also result from barometric condensers.

4. Thermal cracking is employed to break heavy oil fractions into lighter fractions by the application of heat and pressure. Wastewaters result from the overhead accumulator on the fractionator and contain oil fractions, ammonia, phenol, and sulfides. The wastewater is highly alkaline.

5. Catalytic cracking is employed to yield high-quality gasoline and other products. The wastewater (sour water) comes from the steam strippers and overhead accumulators or fractionators. The wastewater is alkaline, high in COD and BOD, and contains oil, sulfides, phenol, and ammonia.

6. Hydrocracking is employed to convert hydrocarbon feedstocks into gasoline and high-quality middle distillates. The process is basically catalytic cracking in the presence of hydrogen. The wastewaters are high in sulfides, phenols, and ammonia.

7. Reforming converts napthas to high-octane gasoline and produces aromatics for petrochemicals or aviation gasoline

through the dehydrocyclization of paraffins. The volume of wastewater is small and will contain some sulfides.

8. Polymerization converts olefin feedstocks to higher-octane polymer gasoline through the use of an acid catalyst. The waste volume is small but high in sulfides, mercaptans, and ammonia.

9. Alkylation is employed for the conversion of normally gaseous hydrocarbons to high-octane motor fuel and is the reaction of an olefin with an aromatic or paraffinic hydrocarbon in the presence of a catalyst. Wastewaters result from the overhead accumulators in the fractionation section, the alkylation reactor, and the caustic wash.

10. Isomerization is a molecular-rearrangement process which produces isobutane and high-octane isomers from normal butane, pentane, and hexane. Wastewaters will contain some phenolics and other oxygen-demanding materials.

11. Solvent refining is a solvent-extraction process which produces high-grade lubricating oil stocks or aromatics. Wastes contain solvent (phenol, glycols, and amines) and oil.

12. Dewaxing removes wax from lube-oil stocks. Wastewaters will contain solvent (methyl ethyl ketone) from leaks and spills.

13. Hydrotreating removes sulfur compounds and color and gum-forming materials from a variety of petroleum fractions by catalytic action with hydrogen. Wastes emanate from overhead accumulators on fractionators, steam strippers, and sour-water stripper bottoms and contain sulfides and ammonia.

14. Deasphalting is a solvent-extraction process for the removal of asphalt or resins from viscous hydrocarbon fractions. Wastewaters will contain small amounts of sulfide, oil, and ammonia.

15. Drying and sweetening removes sulfur compounds, water, and other impurities from gasoline, kerosene jet fuels, etc. The primary waste stream is spent caustic, which may contain high concentrations of phenolics and sulfides.

Wax and grease manufacture may produce small quantities of wastewater containing oil from leaks. Lube-oil finishing removes color-forming and other undesirable materials by clay or acid treatment. Wastewaters are acidic and contain dis-

Table 5.12. Average Waste Flows and Loadings from Petroleum Refineries for Old, Prevalent, and New Technology[a]

Type of Technology	Flow, gal/bbl		BOD, lb/bbl		Phenol, lb/bbl		Sulfide, lb/bbl	
	Avg	Range	Avg	Range	Avg	Range	Avg	Range
Older	250	170–374	0.40	0.31–0.45	0.030	0.028–0.033	0.01	0.008
Typical	100	80–155	0.10	0.08–0.16	0.01	0.009–0.013	0.003	0.0028
Newer	50	20–60	0.05	0.02–0.06	0.005	0.001–0.006	0.003	0.0015
	liters/bbl		g/bbl		g/bbl		g/bbl	
	945	644–1410	181	141–204	13.6	12.7–15	4.5	
	378	301–586	45.4	37.3–72.5	4.5	4.1–5.9	1.4	
	189	76–227	22.7	9.1–27.2	2.3	0.45–2.7	1.4	

[a] From *The Cost of Clean Water* [1].

solved and suspended solids, sulfates, sulfonates, and stable oil emulsions.

The general characteristics of refinery wastes for old, prevalent, and advanced technology are summarized in Table 5.12.

Water Reuse and Waste Reduction   The major water use in a refinery is for cooling (90 percent). Recirculation of cooling water will reduce freshwater intake by greater than 90 percent. The replacement of barometric condensers by surface condensers will reduce both wastewater volumes and loadings because barometric-condenser water contains oil emulsions and requires treatment prior to discharge. The use of air-cooling towers will reduce water consumption by elimination of evaporation losses. In-plant treatment methods to reduce losses to the sewer include sour-water strippers, neutralization and oxidation of spent caustics, ballast water treatment, slop oil recovery, and temperature control.

Waste Treatment   Gravity oil separators (API separation) are employed in all refineries as a first step in waste treatment. Dissolved air flotation, with or without the use of flocculating chemicals, is used by a number of refineries to treat the effluent from a gravity separator. Biological treatment is accomplished by activated sludge, trickling filters, aerated lagoons, or oxidation ponds on oily wastes, sour water, and spent caustic. Pretreatment is required to remove oil and to limit the concentrations of sulfides, mercaptans, and phenols.

Sour-water strippers precede biological treatment and are designed for the removal of sulfides. Depending on the temperature and pH, ammonia, phenols, and cyanide may also be stripped. The anticipated treatment efficiencies are shown in Table 5.13.

Reference

1. *The Cost of Clean Water*, Vol. III, Industrial Waste Profile No. 5, U.S. Department of Interior, FWPCA, Washington, D.C., 1968.

Table 5.13. Efficiency of Oil-Refinery Waste-Treatment Practices[a]

	MPPI[b]	BOD	COD	Phenol	Sulfide S	Suspended Solids
Physical treatment						
API separators	R.W.	5–35[c]	5–30[c]	Reduced	N. A.	10–50
Earthen separators	R.W.	5–50	5–40	Reduced	N. A.	10–85
Air flotation without chemicals	API eff.	5–25	5–20	N. A.	Reduced	10–40
Chemical treatment						
Air flotation without chemicals	API eff.	10–60	10–50	N. A.	Reduced	50–90
Chemical coagulation and precipitation	API eff.	10–70	10–50	N. A.	N. A.	50–90
Biological treatment						
Activated sludge	API eff.	70–95	30–70	65–99	90–99	60–85
Aerated lagoons	API eff.	50–90	25–60	65–99	90–99	0–40
Trickling filters	API eff.	50–90	25–60	65–99	80–99	60–85
Oxidation ponds	API eff.	40–80	20–50	65–99	70–90	20–70
Tertiary treatment						
Activated carbon	SEC.[d] eff.	50–90	50–90	80–99	80–99	N. A.
Ozonation	SEC. eff.	50–90	50–90	80–99	80–99	N. A.

[a] From *The Cost of Clean Water* [1].
[b] MPPI (most probable process influent) indicates the kind and/or extent of prior treatment required for efficient utilization of the specific process under consideration.
[c] BOD and COD from separable oil not included.
[d] Chemical or biological treatment.

PLATING WASTES

Metal-finishing wastes result fom the plating of metal parts into final products. The process involves the stripping, removal of undesirable oxides, cleaning, and plating of the parts. The wastewaters contain acids, alkaline cleaners, grease and oil, and heavy metals such as chromium, zinc, copper, nickel, tin, and cyanides. Wastes emanate primarily from spent plating solutions and from rinse-water dragout.

Several procedures can be employed for waste reduction and water reuse in the plating industry. These include

1. The use of spray rinses and high-pressure fogs instead of running rinses.

2. Use of counterflow rinse systems.

3. Installation of gravity feed, nonoverflowing emergency holding tanks for toxic wastes.

4. Special design of drip pans and a curb area around plating operations.

Chromium wastes can be treated by reduction and precipitation (p. 287) or by ion exchange in which chromic acid is recovered and deionized water reused (p. 235). Cyanide wastes can be treated by the alkaline chlorination process (p. 291). Choice of treatment method will depend upon waste volumes and economics of metal and water recovery.

PULP AND PAPER WASTES

Pulp and paper mills may include several types of operations relative to the raw material used and the preparation of it. Pulp mills include wood preparation, pulping, screening, washing, thickening, and bleaching. Paper mills include stock preparation, paper-machine operation, converting, and finishing.

In a pulp mill, wood is initially moved from the stock pile through a conveyor or log flume to the debarking facilities. The bark is usually mechanically removed and disposed of by incineration, composting, or by-product recovery. The debarked logs are then mechanically chipped in preparation for digestion and pulping.

Pulping is the process of converting wood chips and other fibrous materials into fibers suitable for paper. It is accomplished by mechanical pulping, chemiground wood, or sulfate or sulfite processing.

Mechanical pulping involves grinding the wood in the presence of water and is used for newsprint, low-grade tissue, and paperboard. The chemiground-wood process cooks the whole log. Sulfate or kraft pulping uses alkaline solutions to dissolve liquors and other noncellulosic portions of the wood. The cooking liquors (black liquor) are recovered and reused in most mills. The sulfite process digests the wood in a solution of bisulfite and an excess of SO_2. Recent technology uses ammonium, magnesium, or sodium bisulfite, which are amenable to chemical recovery and water reuse.

The semichemical process uses a mild chemical treatment of softening the wood chips followed by mechanical separation. The pulp yield is higher than the other processes but the strength and flexibility are lower.

Following digestion, the pulp is washed with either diffusers or vacuum filters. Multistage countercurrent washing reduces the water requirements materially.

After washing, the pulp is screened and cleaned to classify the fibers and remove dirt and debris, using screens or centrifugal cleaning devices. This phase of the process may use recycled water. The screened pulp is then thickened to increase its consistency and the water returned to thin fresh stock.

Bleaching is used to brighten the pulp. Ground wood, semichemical, and cold-soda pulps use two-stage hydrosulfite bleaching, while kraft, sulfite, and soda processes use multistage bleaching with various bleaching agents (chlorine, hypochlorite, chlorine dioxide) and caustic extraction with a washing cycle between each stage.

Stock preparation involves treatment of the pulp mechanically and with additives for forming into a sheet on the paper machine. The operations include beating and refining, machine chest mixing, and screening. At this point in the process filters, sizing, wet-strength rinses, and coloring may be added.

The pulp suspension is converted to a paper sheet at the paper machine. This involves three steps: (1) arrangement of

Table 5.15. Average Waste Flows and Loadings from the Pulp and Paper Industry for Old, Prevalent, and New Technology

Type of Technology	Flow, gal/ton Avg	Flow, gal/ton Range	BOD, lb/ton Avg	BOD, lb/ton Range	SS, Avg	SS, Range
Bleached kraft						
Older	110,000	75,000–140,000	200	50–350	200	80–370
Prevalent	45,000	39,000–54,000	120	30–220	170	50–200
Newer	25,000	—	90	—	90	—
Bleached sulfite						
Older	95,000	58,000–170,000	500	350–730	120	50–200
Prevalent	55,000	40,000–70,000	330	235–430	100	40–100
Newer	30,000	10,000–40,000	100	60–300	50	10–70
Bleached kraft						
Older	460	310–580	100	25–175	100	40–185
Prevalent	190	160–220	60	15–110	85	25–100
Newer	105	—	45	—	45	—
Bleached sulfite						
Older	390	240–710	250	175–365	60	25–100
Prevalent	230	170–290	165	118–215	50	20–50
Newer	125	415–170	50	30–150	25	5–35

Table 5.14. Water Reuse and Waste Reduction—Kraft Process[a]

Process	Wastewater	Reuse	Wasteloads Reductions, %	Wastewater Quantities Reductions, %
Wood preparation	Evaporator condensate, bleach pulp washer filtrate	Log flume, debarker, showers	80–90	70–85
Pulping	Recovery condensates	Dilution water, causticizer, screening, de-inking, wood preparation	30 60–90[b]	30 60–90[b]
Screening, washing, and thickening	White water	Makeup water	20–60[c]	20–60[c]
Bleaching			30–80	
Paper machine	White water	Paper machine, stock preparation, bleaching, pulp, washing, and wood preparation	20–70	60–80

[a] From *The Cost of Clean Water* [1].
[b] Liquor recovery.
[c] 60 to 90 percent with multistage countercurrent vacuum filters.

the fibers into a wet web, (2) removal of water by pressing, and (3) heat drying of the sheet. The relatively clean filtrate is recycled to mix with the incoming pulp. The white water from the head end of the machine is returned to a save-all (flotation or filtration), where up to 95 percent of the fibers are recovered. Additives may be added to the paper machine.

Waste and Wastewater Sources   Wastewater results from most of the processes in a pulp and paper mill. The majority of the wastes emanate from the pulping, bleaching, and paper-machine operations. The wastes contain suspended solids (pulp and fiber), BOD from cooking liquor, wood sugars, color from liquor, and other chemicals from the wood preparation processes. The sources and characteristics of water from the individual unit operations are summarized in Fig. 5.7. Mill wastes for old, prevalent, and advanced technology are summarized in Table 5.14.

Water-Reuse and Waste-Recovery Practices   Water may be collected and reused from the log flumes and debarkers, evaporator condensates from the liquor recovery area, washer water from the screens, and bleach-plant washer filtrate and white water from the paper machines. The various areas of water reuse and product recovery are summarized in Table 5.15.

Waste Treatment   Wastewater treatment in the pulp and paper industry generally involves primary and secondary treatment. Suspended solids (pulp and fiber) are removed by sedimentation or flotation. (In some cases flocculant aids may be used.) Biological treatment employs activated sludge for high-level treatment or lagoons (aerated or nonaerated). Nutrients (nitrogen and phosphorus) must be added. Biological treatment will not remove color and some finely divided suspended solids. Sludges from pulp and paper mill wastes are dewatered by centrifugation or vacuum filtration. The efficiencies of the various wastewater-treatment processes in use today are shown in Table 5.16.

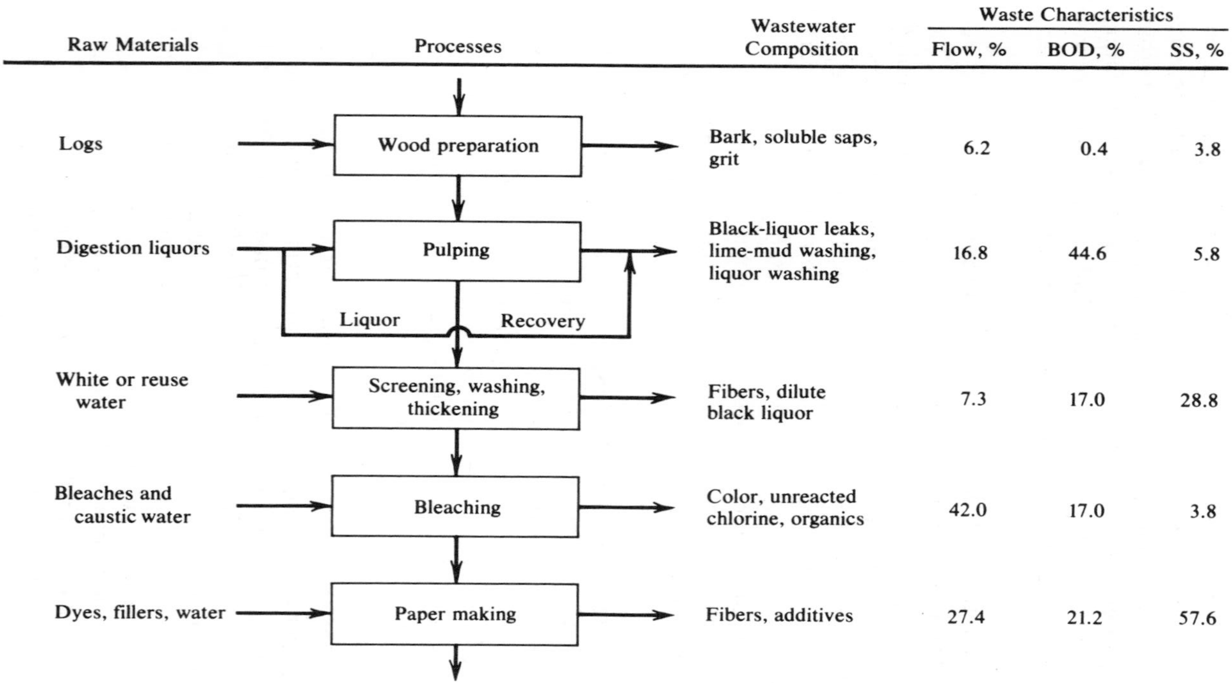
Raw Materials
Processes
Wastewater Composition
Waste Characteristics
Flow, %
BOD, %
SS, %
Logs
Wood preparation
Bark, soluble saps, grit
6.2
0.4
3.8
Digestion liquors
Pulping
Liquor
Recovery
Black-liquor leaks, lime-mud washing, liquor washing
16.8
44.6
5.8
White or reuse water
Screening, washing, thickening
Fibers, dilute black liquor
7.3
17.0
28.8
Bleaches and caustic water
Bleaching
Color, unreacted chlorine, organics
42.0
17.0
3.8
Dyes, fillers, water
Paper making
Fibers, additives
27.4
21.2
57.6

Figure 5.7 Wastewater characteristics from the kraft process.

Table 5.16. Waste-Treatment Removal Efficiencies[a]

	SS, %	BOD, %	COD, %	Color, %
Sedimentation	60–90	10–40	10–30	< 10
Dissolved air				
flotation	70–95	20–50	10–40	
Stabilization basin	< 90	30–50	—	< 10
Aerated lagoon	—	40–85	30–60	< 10
Activated sludge	70–98	< 95	< 70	< 30

[a] From *The Cost of Clean Water* [*1*].

Reference

1. *The Cost of Clean Water*, Vol. III, Industrial Waste Profile No. 3, U.S. Department of Interior, FWPCA, Washington, D.C., 1968.

STEEL-MILL WASTES

General There are several processes in steel making which result in wastewater. Pig iron is produced in the blast furnace from iron ore, coke, and limestone as raw materials. Steel is produced in one of several processes, including the open hearth furnace, the basic oxygen furnace, the electric furnace, and the Bessemer converter.

The molten steel is cast into ingots which go to the rolling mills where semifinished shapes are formed. These are further reduced by rolling into plates, strips, bars, and other forms. During the rolling steps cleaning is required to remove oxides, films, oils, and dirt. This may include water washing, shot blasting, pickling, and the use of detergents and solvent.

Some plants provide finishing operations, including galvanizing, tin plating, and coating.

Coke plants are used to provide coke for the iron-making process. Bituminous coal is baked in ovens in the absence of air in which volatile material is driven off, leaving coke as a residue. Some of the volatile materials are recovered and some remain in the wastewaters.

A typical integrated steel mill has one or more blast furnaces with daily capacities of 1000 tons of iron. Steel making is accomplished in oxygen-lanced open-hearth furnaces. Steel pickling uses high-speed strip picklers with sulfuric acid. By-product coke ovens use ammonia liquor for quenching with maximum by-product and sometimes ammonia recovery.

Waste and Wastewater Sources   Wastes from blast furnace operation is water used to wash the exit gases from the furnace. These wastes may contain from 1000 to 10,000 mg/l of suspended solids at 600 to 5000 gpm per furnace. Fifty percent of the particles are finer than 10 μ in diameter and have a specific gravity of 3.5. Cyanides, phenols, and ammonia will also be present in the wastewater.

Casting machines and hot rolling mills produce wastewater containing saponified oils, dirt, alkali, and solvent residues. Cleaning operations produce pickle liquor which contains acids and iron salts of these acids. The pickling wastes may contain 5 to 9 percent free acid and 13 to 16 percent iron salts.

Steel-finishing operations produce wastewaters containing emulsified oils and suspended solids. Plating operations will contain soluble metals. The wastes may contain 200 mg/l of oil.

Coke-plant wastes will contain phenols, ammonia, ammonium salts, cyanides, chlorides, and sulfates.

The wastewater characteristics from a steel mill are summarized in Table 5.17 for various levels of technology. The newer technologies tend to produce greater waterborne waste loads because lighter products are becoming more predominant and the wastes generated are generally in proportion to the surface area of steel exposed during rolling and finishing operations and to the relative gas–liquid interfacial areas in iron- and steel-making operations.

Water-Reuse and Waste-Recovery Practices
Waste reduction in steel mill operations has been accomplished by the following methods:

1. Recirculation of gas wash water in blast furnaces with about 20 percent blowdown and recovery of mill scale.

2. The use of magnetic separators in cold rolling mills.

3. The use of HCl instead of H_2SO_4 reduces waste acid by 20 percent.

Table 5.17   Steel Mill Wastewater Characteristics based on Quantity Produced per Ingot Ton Per Day[a]

	Old	Prevalent	New
Flow, gal/ton	9860	10,000	13,750
Suspended solids, lb/ton	103	125	184
Phenols, lb/ton	0.069	0.064	0.064
Cyanides, lb/ton	0.029	0.028	0.031
Fluorides, lb/ton	0.033	0.031	0.031
Ammonia, lb/ton	0.082	0.078	0.078
Lube oils, lb/ton	3.08	2.72	2.37
Free acids, lb/ton	3.03	3.54	3.40
Emulsions, lb/ton	0.332	0.414	1.17
Soluble metals, lb/ton	—	0.079	0.082
	Metric Units		
Flow, m^3/metric ton	41	42	57
Suspended solids, kg/metric ton	52	63	92
Phenols, kg/metric ton	0.035	0.032	0.032
Cyanides, kg/metric ton	0.015	0.014	0.015
Ammonia, kg/metric ton	0.041	0.039	0.039
Lube oils, kg/metric ton	1.54	1.36	1.18
Free acid, kg/metric ton	1.51	1.77	1.70
Emulsions, kg/metric ton	0.161	0.207	0.58
Soluble metals, kg/metric ton	—	0.040	0.041

[a] From *The Cost of Clean Water* [1].

4. Shot blasting to replace acid pickling.

5. The use of pickling-liquor regeneration processes.

6. The use of benzol plant wastes and still wastes for coke-plant quenching and cooling-water recirculation reduces phenol cyanide and ammonia by 80 to 98 percent.

7. Production of ammonium sulfate by reaction of the ammonia with sulfuric acid.

8. Phenol recovery by vapor recirculation or solvent extraction.

Waste-Treatment Practices   Suspended solids present in blast furnace and rolling mill wastes are removed by sedimentation. Removal efficiencies (yielding effluent suspended solids contents of less than 50 mg/l) can be increased by the addition of coagulant aids. Coagulation and sedimentation or flotation is used for the removal of emulsified and free oils. Phenols and cyanides are removed by biological oxidation. Plating wastes can be treated by ion exchange or reduction and precipitation. Concentrated wastes can be disposed of in deep wells. Pickling liquor not regenerated is neutralized prior to discharge. The various treatment processes and normal removal efficiencies are summarized in Table 5.18.

Table 5.18   Normal Removal Efficiencies for
Steel Mill Waste Treatment

	Percent Removal	
Process	Suspended Solids	Lube Oils
Sedimentation	90–94	20
Coagulation	95–98	80
Recirculation and sedimentation	96–98	60
Recirculation and coagulation	98–99	90
Magnetic separators	80	

Reference
1. *The Cost of Clean Water*, Vol. III, Industrial Waste Profile No. 1, U.S. Department of Interior, FWPCA, Washington, D.C., 1968.

TANNERY WASTES

General Leather tanning and finishing consists of several steps which convert animal hides to finished leather.

1. Hide trimming and storage—no liquid waste is generated.

2. Soaking and washing—removal of dirt, salt, manure, blood, etc.; lime and sodium sulfide might be added to solublize coagulated protein, disperse fats, etc.

3. Defleshing—mechanical process to remove fleshing from hides; fleshings recovered and sold as a by-product.

4. Dehairing—removal of hair and open-fiber structure to remove undesirable protein; process used depends on hair-recovery practice. Most common chemicals used are lime (4 to 12 percent by weight of the hides) and sodium sulfide and sodium sulfhydrate (0.1 to 3.0 percent)(higher strengths for hair dissolution). Dimethylamine sulfate can replace sulfide and reduce the BOD load up to 50 percent.

5. Lime splitting—mechanical process of splitting hides, no liquid wastewater.

6. Bating—hide soaking in a solution of ammonium salts and enzymes to prepare hides for tanning; remove protein-decomposition products.

7. Pickling—treatment of hides with sulfuric acid and salt to prevent precipitation of chromium during tanning; not required for vegetable tanning.

8. Tanning—conversion of fibers in hides to leather. Chrome tanning uses chrome sulfate which reacts with the skin protein and is primarily used for light leathers; vegetable tanning is used for heavier leathers. Reuse and recovery is practiced in all vegetable tanneries. The spent tan liquors are low in volume and high in pollutional load.

9. Finishing—processes include vegetable retaining of chrome-tanned leathers, dyeing and coloring, and bleaching.

Wastewater Sources   Wastewaters from tanning operations contain free lime and sulfides; are high in pH, BOD, and suspended solids; and contain trivalent chromium in the case of chrome tanning.

Wastewaters result from two primary areas, the beam house (steps 1 to 5) and the tanhouse (steps 6 to 9). As much as 80 percent of the waste volume and 90 percent of the suspended solids originate in the beam house. The major portion of the BOD also originates in the beam house, the percentage being higher in the case of chrome tanning. The relative percentages of wastewater and pollutants from the various process steps is shown in Fig. 5.8 for tanneries with prevalent or new technology.

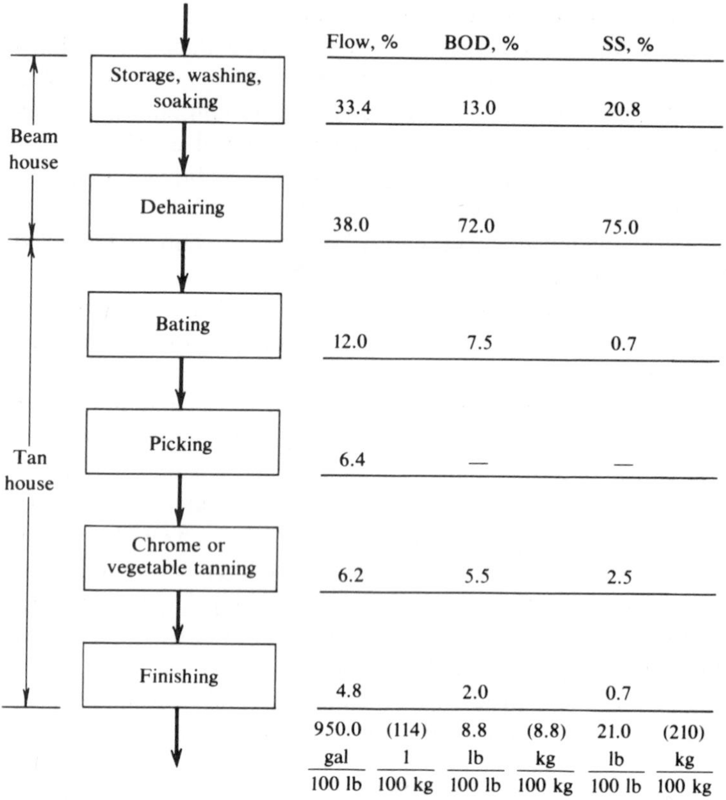

	Flow, %	BOD, %	SS, %
Storage, washing, soaking	33.4	13.0	20.8
Dehairing	38.0	72.0	75.0
Bating	12.0	7.5	0.7
Picking	6.4	—	—
Chrome or vegetable tanning	6.2	5.5	2.5
Finishing	4.8	2.0	0.7

Beam house

Tan house

$$\frac{950.0 \quad (114)}{\frac{\text{gal}}{100\ \text{lb}} \quad \frac{1}{100\ \text{kg}}} \qquad \frac{8.8 \quad (8.8)}{\frac{\text{lb}}{100\ \text{lb}} \quad \frac{\text{kg}}{100\ \text{kg}}} \qquad \frac{21.0 \quad (210)}{\frac{\text{lb}}{100\ \text{lb}} \quad \frac{\text{kg}}{100\ \text{kg}}}$$

Figure 5.8 Relative percentage of wastewater and pollutants from a tannery.

Wastewater Characteristics Wastewater volumes and pollutional loads from tannery operation have been summarized per pound of leather processed[1].

Technology	BOD, lb (kg)	SS, lb (kg)	TDS, lb (kg)	Volume gal (l)
Older	0.0916 (0.0416)	0.260 (0.118)	0.380 (0.172)	10.5 (3.98)
Prevalent– newer	0.0883 (0.0400)	0.250 (0.113)	0.350 (0.159)	9.5 (3.50)

The statistical variation in wastewater characteristics per 100 lb (45.4 kg) of hides processed based on 50 datum points is summarized in Table 5.19. The wastewater volume will vary depending on the type of tanning used. For example, the average reported wastewater flows for chrome tanning, vegetable tanning, and a mixed tannery are 1630, 670, and 980 gal/100 lb of hides (13,600, 5580, 8160 l/100 kg of hides), respectively.

Table 5.19   Characteristics of Tannery Wastewaters

	Probability of Occurrence		
	Mean	16%	84%
Flow			
gal/unit	660	440	1010
liters/unit	250	166	383
BOD			
lb/unit	6.2	2.0	10.5
kg/unit	2.8	0.91	4.8
SS			
lb/unit	13.0	5.0	21.0
kg/unit	5.9	2.3	9.5

Water-Reuse, Product-Recovery, and Waste-Reduction Practices Recently several procedures and process changes have become available to reduce both the volume and strength of tannery wastewaters. The use of rotating drums on a semicontinuous basis to replace batch tanks with paddles will yield a considerable reduction in waste volume. The substitution of dimethylamine sulfate for sulfides will reduce the BOD loading and eliminate sulfide problems in the wastewater. Reuse and recovery of vegetable tanning liquors yields a substantial BOD reduction. The addition of calcium or sodium formate to chrome tanning liquors lowers the required chrome concentration and improves the chrome takeup by the hide, thus lowering the chrome concentration in the spent tan liquor dumped to the sewer. The use of syntans (synthetic tanning materials) in lieu of natural vegetable extracts results in considerable color reduction. Sodium pentachlorophenate, which is colorless, can be used as a fungicide instead of p-nitrophenol, which is highly colored.

Waste-Treatment Practices   It is estimated that in the near future 80 percent of the tannery wastes in the United States will be discharged to municipal sewers. Hair, grease, and fleshings should be removed at the tannery to avoid sewer clogging. In some cases the high sulfide content can lead to corrosion of the crowns of concrete sewers. High treatment levels are possible, but equalization should be considered if the ratio of tannery waste to municipal sewage is very high because the high pH, chromium, and sulfide content can shock load and upset secondary municipal facilities.

Separate treatment of tannery wastes can produce varying degrees of effluent quality, depending on the combination of treatments employed. The ranges of reported performance are summarized in Table 5.20.

Table 5.20.   Efficiency of Treatment of Tannery Wastes[a]

Process	Removal Efficiency, %		
	BOD	SS	Sulfides
Screening	—	5–10	—
Sedimentation	25–62	69–96	5–20
Coagulation	41–70	70–97	14–50
Lagoons	70	80	—
Activated sludge	85–95	80–95	75–100

[a] From *The Cost of Clean Water* [1].

Reference
1. *The Cost of Clean Water*, Vol. III, Industrial Waste Profile No. 7, U.S. Department of Interior, FWPCA, Washington, D.C., 1968.

TEXTILE-PRODUCT WASTES

The three principal product categories in the textile industry are wool weaving and finishing, cotton textile finishing, and synthetic textile finishing. There are four basic processes in the finishing of textiles common to each category: (1) scouring, (2) dyeing and/or printing, (3) bleaching, and (4) special finishing.

Wool is generally scoured, washed, and dyed before being woven into cloth; cotton and synthetic fibers are generally woven into cloth before any finishing operations are performed.

Desizing and scouring removes natural impurities and chemical additives. In the wool industry, scouring uses a solution of soap or synthetic detergent, alkali, and water, and produces an alkaline wastewater containing dirt and grease approximately equal to the original fiber weight.

Desizing and scouring of cotton uses enzymes, water, and alkali to remove starch and other sizing agents previously added to the cloth, as well as natural impurities such as wax, pectins, and dirt. Desizing and scouring may be separate or combined operations. The wastewaters are alkaline and high in BOD.

Synthetic-fiber scouring wastes are low in volume and concentration and may contain weak alkali, antistatic agents, lubricants, soap or detergents, and sizing.

Dyes may include direct acid, fiber reactive, vat, basic napthol, sulfur, and acetate, depending on characteristics of the fiber and the finish desired. Synthetic fibers may use special carriers. Printing uses dye pastes followed by chemical treatment and washing.

Bleaching is done in batch or continuous processes: cotton using hypochlorite or hydrogen peroxide, wool using hydrogen peroxide and acid, and synthetic fibers using hydrogen peroxide, sodium chlorite, peracetic acid, sodium chloride, or other chemicals. The wastewaters are acidic, may be toxic, but are low in BOD.

Special finishing includes waterproofing, brightening, anti-wrinkling, and mercerizing for cotton. These processes usually employ a chemical treatment followed by washing for the removal of residues. The approximate waste load produced by each of these processes is shown in Table 5.21.

Waste volumes and characteristics from the textile industry will vary markedly depending both on the type of fiber and the processing procedures employed. The general character-

Table 5.21 Percent of BOD and Solids

Process	Wool	Cotton	Synthetic
Scouring	50–75	50–75	<50
Dyeing	3–10	15–35	5–80
Bleaching	<5	<5	<5
Special finishing	5–15	5–15	5–15

istics are summarized in Table 5.22. The more advanced technology employs continuous processes with countercurrent washing, better grease recovery, and substitution of less-pollution-possessing chemicals.

Waste-reduction and water-reuse practices have resulted in substantial reduction of pollutional loads, as indicated by the level of technology in Table 5.22. Some of the more common procedures employed are summarized in Table 5.23.

The treatment of textile wastes usually employs biological treatment preceded by pretreatment for neutralization, grease

Table 5.22 Average Wastewater Characteristics
from the Textile-Finishing Industry[a]
(per 1000 lb of cloth produced)

Process	Technology	Vol, 1000 gal	BOD, lb	SS, lb	TDS, lb
Wool	Old	73.7	450	—	—
	Prevalent	63.0	300	—	—
	New	62.0	50	—	—
Cotton	Old	50.0	170	80	245
	Prevalent	38.0	155	70	205
	New	35.0	140	62	187
Synthetic	Rayon	3–7	20–40	20–90	20–500
	Acetate	7–11	40–50	20–60	20–300
	Nylon	12–18	35–55	20–40	20–300
	Acrylic	21–29	100–150	25–150	25–400
	Polyester	8–16	120–250	30–160	30–600
		Metric Units			
Wool	Old	614			
	Prevalent	525			
	New	516			
Cotton	Old	416			
	Prevalent	316			
	New	292			
Synthetic	Rayon	25–58			
	Acetate	58–92			
	Nylon	100–150			
	Acrylic	175–241			
	Polyester	67–133			

[a] From The Cost of Clean Water [1].

Table 5.23. Waste-Reduction Practices[a]

Reduction, %		
Wool scouring	12	Substitution of synthetic detergent for soaps
	96	Use of solvents
Washing (wool)	80	Substitution of synthetic detergents for soaps
Desizing (cotton)	70	Substitution of sodium carboxymethylcellulose (CMC)
Continuous scouring (cotton)	20	Use of less NaOH
Mercerizing (cotton)	60	NaOH recovery
Dyeing (cotton)	80	Synthetic detergents used in wash after dyeing

[a] From *The Cost of Clean Water* [1].

removal, and other purposes. The various treatment processes and average efficiencies are summarized in Table 5.24.

Table 5.24 Waste-Treatment Processes and Efficiencies[a]
(% reduction)

Treatment Process	BOD	Color	Alkalinity	SS
Sedimentation	$30-50^b$	$10-50^b$	$10-20^b$	$50-65^b$
	$5-15^{c,d}$	—	—	$15-60^{c,d}$
Chemical	$20-85^{b,e}$	$<75^b$	—	—
Coagulation	$25-60^c$	—	—	$<90^c$
Lagoons	$<85^b$	$<30^b$	$<20^b$	$<70^b$
	$<80^c$	—	—	$<80^c$
Aerated lagoons	—			
	$<95^{c,d}$	—	—	$<95^{c,d}$
Activated sludge	$<90^b$	$<30^b$	$<30^b$	$<95^b$
	$<95^{c,d}$	—	—	$<95^{c,d}$

[a] From *The Cost of Clean Water* [1]. [b] Wool finishing.
[c] Cotton finishing. [d] Synthetic fibers.
[e] H_2SO_4 + alum or H_2SO_4 + $FeCl_2$ or recovery by acid cracking, centrifugation, or evaporation.

Reference

1. *The Cost of Clean Water*, Vol. III, Industrial Waste Profile No. 4, U.S. Department of Interior, FWPCA, Washington, D.C., 1968.

6

Wastewater-Treatment Processes

There are a large number of wastewater-treatment processes in use whose application is related both to the characteristics of the waste and the degree of treatment required. The various processes as a treatment sequence and substitution diagram are shown in Fig. 6.1. Pretreatment or primary treatment is used for the removal of floating and suspended solids and oils, neutralization and equalization, and to prepare the wastewater for subsequent treatment or discharge to a receiving water. General considerations in pretreatment or primary treatment for secondary biological treatment are summarized in Table 6.1 (more specific details are given on p. 189). The processes gen-

Table 6.1. Pretreatment or Primary Treatment Requirements for Biological Processes

Characteristics	Treatment
Suspended solids	Lagooning, sedimentation, and flotation
Oil or grease	Skimming tank or separator
Heavy metals	Precipitation or ion exchange
Alkalinity	Neutralization for excessive alkalinity
Acidity	Neutralization
Sulfides	Precipitation or stripping
BOD loading	Equalization

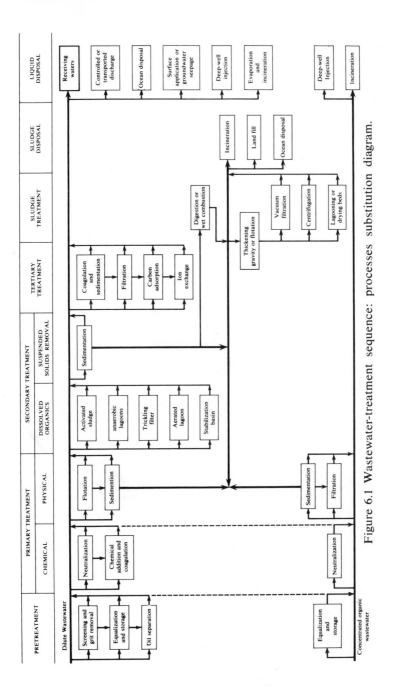

Figure 6.1 Wastewater-treatment sequence: processes substitution diagram.

erally applicable to the removal of specific pollutants are shown in Table 6.2. The characteristics of the effluent from

Table 6.2.   Processes Applicable to Wastewater Treatment

Pollutant	Processes
Biodegradable organics (BOD)	Aerobic biological (activated sludge), aerated lagoons, trickling filters, stabilization basins, anaerobic biological (lagoons, anaerobic contact), deep-well disposal
Suspended solids (SS)	Sedimentation, flotation, screening
Refractory organics (COD,TOC)	Carbon adsorption, deep-well disposal
Nitrogen	Maturation ponds, ammonia stripping, nitrification and denitrification, ion exchange
Phosphorus	Lime precipitation; Al or Fe precipitation, biological coprecipitation, ion exchange
Heavy metals	Ion exchange, chemical precipitation
Dissolved inorganic solids	Ion exchange, reverse osmosis, electrodialysis

these processes under optimum operation are summarized in Table 6.3.

The selection of a wastewater-treatment process or a combination of processes will depend upon:

1. The characteristics of the wastewater.
2. The required effluent quality.
3. The costs and availability of land.
4. The future upgrading of water-quality standards.

Each of the processes and the applicable design criteria are discussed in detail in subsequent sections of this book.

Table 6.3. Maximum Effluent Quality Attainable from Waste-Treatment Processes

Process	BOD	COD	SS	N	P	TDS
Sedimentation, % removal	10–30	—	50–90	—	—	—
Flotation,[a] % removal	10–50	[b]	70–95	—	[c]	—
Activated sludge, mg/l	<25	—	<20	[c]	—	—
Aerated lagoons, mg/l	<50	—	>50	—	—	—
Anaerobic ponds, mg/l	>100	—	<100	—	—	—
Deep-well disposal	Total disposal of waste					
Carbon adsorption, mg/l	<2	<10	<1	—	—	—
Ammonia stripping, % removal	—	—	—	>95	—	—
Denitrification and nitrification, mg/l	<10	—	—	<5	—	—
Chemical precipitation, mg/l	—	—	<10	—	<1	—
Ion exchange, mg/l	—	—	<1	[d]	[d]	[d]

[a] Higher removals are attained when coagulating chemicals are used.

[b] COD_{inf} - [BOD_u (Removed)/0.9].

[c] N_{inf} - 0.12 (excess biological sludge), lb; P_{inf} - 0.026 (excess biological sludge), lb.

[d] Depends on resin used, molecular state, and efficiency desired.

7

Pretreatment and Primary Treatment

Pretreatment and primary treatment are employed primarily for the removal of floating materials and suspended solids and for conditioning the wastewater for discharge to waters of low classification or secondary treatment through neutralization and/or equalization. The various processes used in pretreatment and primary treatment are shown on page 109. The primary-treatment requirements for secondary treatment are summarized on page 189.

SCREENING

Screening is employed for the removal of large solids prior to other treatment processes. In municipal sewage treatment, screens are usually provided at the head end of the plant for the removal of coarse materials. The screens consist of coarse bars or racks with $1\frac{1}{2}$ - to $2\frac{1}{2}$ -in. (3.8 to 6.4-cm) openings and may be either mechanically or manually cleaned, as shown in Fig. 7.1. The quantity of screenings which might be expected from municipal sewage are discussed on page 244.

Screens for industrial waste treatment are usually of the rotary, vibrating, or eccentric type and are widely used in the canning, brewing, and pulp and paper industries.

SEDIMENTATION

Sedimentation is employed for the removal of suspended solids from wastewaters. The process can be considered in

Figure 7.1 Mechanical sewage screen. (Courtesy of Rex Chainbelt, Inc.)

three basic classifications, depending on the nature of the solids present in the suspension: discrete, flocculent, and zone settling. In discrete settling, the particle maintains its individuality and does not change in size, shape, or density during the settling process. Discrete settling is observed with suspensions of grit (grit chambers), fly ash, and coal. Flocculent settling occurs when the particles agglomerate during the settling period, resulting in a change in size and settling rate. Examples include domestic sewage and pulp and paper wastes. Zone settling involves a flocculated suspension which forms a lattice structure and settles as a mass, exhibiting a distinct interface during the settling process. Alum flocs and activated sludge usually exhibit zone settling.

Discrete Settling   The settling velocity of discrete particles is related to gravity and viscous forces and is defined by the relationship

$$v = \sqrt{\frac{4g\,(\rho_s - \rho_L)\,D}{3C_D}} \tag{1}$$

where ρ_s = specific gravity of the particle
$\quad\;\rho_L$ = specific gravity of the liquid
$\quad\;D$ = diameter of the particle
$\quad\;C_D$ = drag coefficient

The drag coefficient is related to the Reynolds number as shown in Fig. 7.2 for spherical particles and encompasses three regions. Many of the solids-separation problems encountered in sewage and waste treatment are defined by Stokes' law.

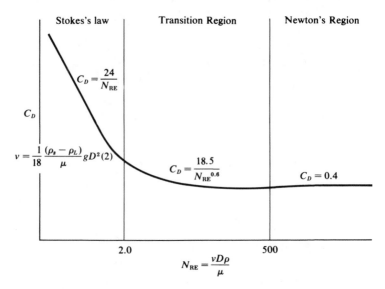

Figure 7.2 Settling characteristics of discrete particles.

The ideal tank concept was developed by Hazen [1] and Camp [2] to define relationships applicable to the design of sedimentation basins. This concept is based on the premises that particles entering a tank are uniformly distributed across the influent cross section and that a particle is considered removed when it reaches the bottom of the tank. The settling velocity of any particle can be related to the overflow rate:

$$V = \frac{D}{t} = \frac{D \cdot A}{A \cdot t} = \frac{V/t}{A} = \frac{Q}{A}$$

Referring to Fig. 7.3 and the premises established for the ideal settling tank, all particles with settling velocities greater than V_0 will be completely removed and particles with settling velocities less than V_0 will be removed in the ratio V/V_0 or h/h_0.

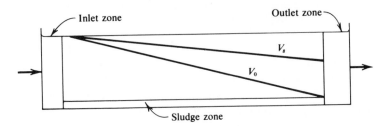

Figure 7.3 Discrete particle settling.

The removal of discrete particles is related only to overflow rate. The removal of particles from suspensions with a wide range of particle sizes can be computed by graphical integration from the relationship

$$\text{Removal} = (1 - C_0) + \frac{1}{V_0}\int_0^{C_0} V\, dc$$

in which C_0 is the fraction of particles with a settling velocity equal to or less than V_0 as shown in Fig. 7.4.

Since settling basins are subject to turbulence, short-circuiting, and velocity gradients, a correction must be made to the ideal settling tank. Dobbins [3] and Camp [2] have developed

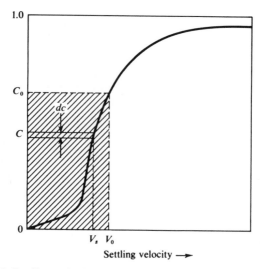

Figure 7.4 Settling-velocity analysis curve for suspension of nonflocculating particles.

a correction for turbulence. Alternatively, depending on the hydraulic characteristics of the settling basin, the overflow rate will be decreased by a factor of 1.25 to 1.75 and the detention period increased by a factor of 1.5 to 2.0 [4].

Scour occurs when the flow through velocity is sufficient to resuspend previously settled particles. Scour is usually not a problem in large settling tanks, but can occur in grit chambers and narrow channels.

Flocculent Settling   Flocculent settling occurs when the settling velocity of the particles increases as it settles through the tank depth, owing to coalescence with other particles, thereby increasing the settling rate and yielding a curvilinear settling path, as shown in Fig. 7.5.

When flocculation occurs, both overflow rate and detention time become significant factors for design.

Since a mathematical analysis is not possible in the case of flocculent suspensions, a laboratory settling analysis is required to establish the necessary parameters [5].

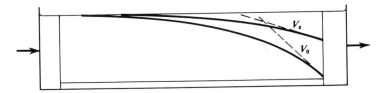

Figure 7.5 Flocculant particle settling.

The degree of flocculation will be influenced by the initial concentration of suspended solids, so that the anticipated range of suspended solids in the wastewater should be considered in the design. Correction for turbulence, short-circuiting, and other conditions must be applied to the relationships developed from the laboratory study.

The suspended solids and BOD removal from domestic sewage and from pulp and paper mill wastes are shown in Fig. 7.6.

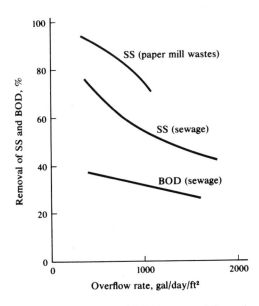

Figure 7.6 Suspended solids and BOD removal from domestic sewage and paper mill wastes.

Zone Settling   Zone settling is characterized by activated sludge and flocculated chemical suspensions when the
concentration of solids exceeds approximately 500 mg/l.
The floc particles adhere together and the mass settles as a
blanket, forming a distinct interface between the floc and the
supernatent. The settling process is distinguished by three
zones, as shown in Fig. 7.7. Initially, all the sludge is at a
uniform concentration.

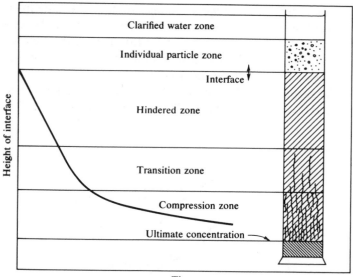

Figure 7.7 Settling zones.

During the initial settling period, the sludge settles at a uniform velocity. The settling rate is a function of the initial solids
concentration. As settling proceeds, the collapsed solids on the
bottom of the settling unit buildup at a constant rate. A zone of
transition results, through which the settling velocity decreases
as a result of an increasing concentration of solids. The concentration of solids in the zone settling layer remains constant
until the settling interface approaches the rising layers of col-

lapsed solids and a transition zone occurs. Through the transition zone, the settling velocity will decrease, owing to the increasing density and viscosity of the suspension surrounding the particles. When the rising layer of settled solids reaches the interface, a compression zone occurs.

In the separation of flocculent suspensions, both clarification of the liquid overflow and thickening of the sludge underflow are involved. The overflow rate for clarification requires that the average rise velocity of the liquid overflowing the tank be less than the zone settling velocity of the suspension. The tank-surface-area requirements for thickening the underflow to a desired concentration level are related to the solids loading to the unit and are usually expressed in terms of a mass loading, lb of solids/ft²/day (kg/m²/day), or a unit area, ft²/lb of solids/day (m²/kg/day).

Clarifiers Clarifiers may either be rectangular or circular. In most rectangular clarifiers, scraper flights extending the width of the tank move the settled sludge toward the inlet end of the tank at a speed of about 1 ft/min. Some designs move the sludge toward the effluent end of the tank, corresponding to the direction of flow of the density current. A typical unit is shown in Fig. 7.8.

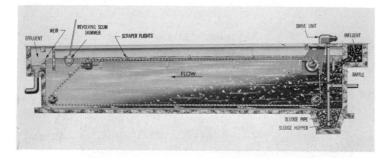

Figure 7.8 Rectangular clarifier. (Courtesy of Chain Belt Company.)

Circular clarifiers may employ either a center feed well or a peripheral inlet. The tank can be designed for center sludge withdrawal or vacuum withdrawal over the entire tank bottom.

Center sludge withdrawal requires a minimum bottom slope of 1 in./ft (8.5 cm/m). The flow of sludge to the center well is largely hydraulically motivated by the collection mechanism, which serves to overcome inertia and avoid sludge adherence to the tank bottom. The vacuum drawoff is particularly adapt-

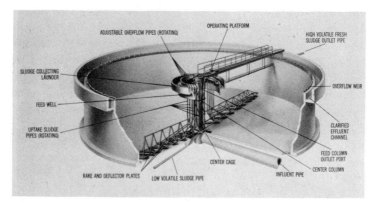

Figure 7.9 Circular clarifier. (Courtesy of Dorr Oliver Incorporated.)

able to secondary clarification and thickening of activated sludge. These units are shown in Fig. 7.9 and 7.10.

The mechanisms can be of the plow type or the rotary-hoe type. The plow-type mechanism employs staggered plows at-

Figure 7.10 Clarifier with suction sludge drawoff. (Courtesy of Chain Belt Company.)

tached to two opposing arms which move about 10 ft/min (3 m/min). The rotary-hoe mechanism consists of a series of short scrapers suspended from a rotating supporting bridge on endless chains which make contact with the tank bottom at the periphery and move to the center of the tank.

An inlet device is designed to distribute the flow across the width and depth of the settling tank. The outlet device is likewise designed to collect the effluent uniformly at the outlet end of the tank. Inlet and outlets of good design will reduce the short-circuiting characteristics of the tank. Increased weir length can be provided by extending the effluent channels back into the basin or by providing multiple effluent channels. In circular basins, inboard or radial weirs will ensure low takeoff velocities. Relocation of weirs is sometimes necessary to minimize solids carryover induced by density currents resulting in upwelling swells of sludge at the end of the settling tank.

OIL SEPARATION

In an oil separator, free oil floats to the surface of the tank, where it is skimmed off. The principles discussed on page 114 apply except that the lighter-water oil globules rise through the liquid. The design of oil separators as developed by the American Petroleum Institute [6] is based on the removal of free oil globules larger than 0.015 cm in diameter. The Reynolds number is less than 0.5, so Stokes' law applies. A procedure that considers short-circuiting and turbulence has been developed [6]. An oil separator is shown in Fig. 7.11. Design improvement such as the addition of inclined parallel plates to serve as collecting sufaces for oil globules and modifications to the inlet structure have improved separator performance.

PROCEDURE FOR THE DESIGN OF A PRIMARY SETTLING TANK

Data to Be Collected   Flow and characteristics of the wastes, such as suspended solids concentration, BOD (if organic), and temperature.

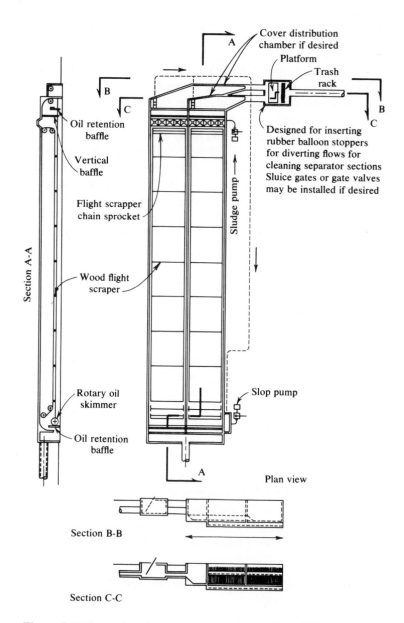

Figure 7.11 Example of general arrangement for API separator. (Courtesy of the American Petroleum Institute.)

Design Procedure

1. Develop the settling rate–time relationship curves for at least three different concentrations of suspended solids to cover the expected range of fluctuation in suspended solids concentrations (Fig. 7.12).

2. The overflow rates and detention times for various percent removals are computed from the above curves as follows. The overflow rate or the settling velocity V_0 is the effective depth (6 ft) divided by the time required for a given percent to settle through the effective depth. All particles having a settling velocity equal to or greater than V_0 will be removed in an ideal basin. Particles with a lesser settling velocity, V, will be removed in proportion V/V_0. For a given settling depth and a detention period t (min), a certain percentage of the suspended solids is removed completely (Fig. 7.12a). Particles in each additional 10 percent range will be removed in the proportion V/V_0 or in proportion to the average depth settled to the total settling depth. Each subsequent percentage range is computed in a similar manner and the total removal determined as follows:

$$\text{total percent removal} = x + \frac{d_1}{d_0}(10) + \frac{d_2}{d_0}(10) + \frac{d_3}{d_0}(10)$$

Overflow rates [as gal/ft²/day (m³/m²/day)] are computed from settling velocities.

3. Overflow rates vs. percent suspended solids removal (Fig. 7.12b) and retention time vs. percent suspended solids removal (Fig. 7.12c) are developed.

4. For a required percent suspended solids removal, the overflow rate and the detention time for a given initial suspended solids concentration are selected.

5. For prototype design, the overflow rate is decreased by a factor of 1.25 to 1.75 and the detention time increased by a factor of 1.5 to 2.0 to account for the effects of turbulence, short-circuiting, and inlet and outlet losses.

6. Compute the surface area and depth.

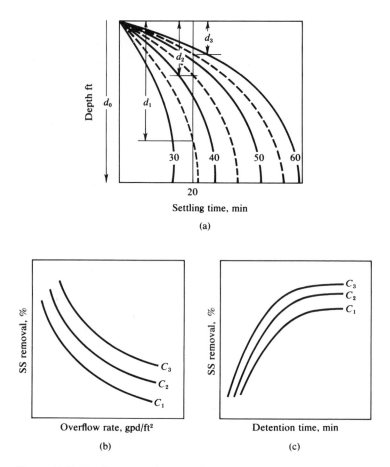

Figure 7.12 Design procedure for flocculent settling. (a) Percent SS removed vs. time and depth; (b) overflow rate vs. percent removal; (c) detention time vs. percent removal.

FLOTATION

Flotation is a process in which the waste liquid or a recycle is pressurized to 2 to 4 atm in the presence of air and then released to atmospheric pressure in the flotation unit. Under turbulent mixing conditions in the tank, air bubbles come out of solution and attach themselves to and become enmeshed in the suspended particles present in the waste, causing them to rise to the surface of the tank. The concentrated suspended solids are scraped from the tank surface and the clarified liquid withdrawn from near the bottom.

The basic components of a flotation system are a pressurizing pump, retention tank to provide air–liquid contact, a pressure-reducing valve, and the flotation tank. A typical flow diagram is shown in Fig. 7.13.

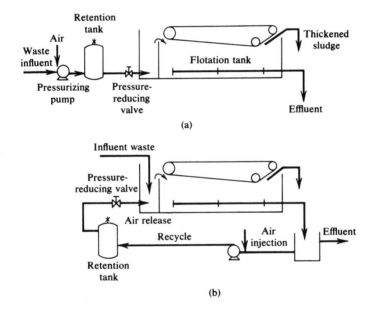

(a)

(b)

Figure 7.13 Schematic representation of flotation systems. (a) Flotation system without recirculation; (b) flotation system with recirculation. (From Eckenfelder [7].)

Effective solids–liquid separation and concentration of the separate solids depends upon releasing sufficient air bubbles relative to the solids present in the waste. This is expressed as an air-to-solids ratio (A/S) as lb of air released/lb of solids (kg/kg). The flotation unit itself is designed based on overflow rate or solids loading (see p. 119). A flotation unit is shown in Fig. 7.14.

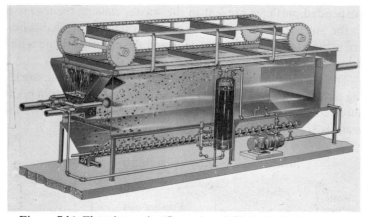

Figure 7.14 Flotation unit. (Courtesy of Chain Belt Company.)

PROCEDURE FOR THE DESIGN OF A FLOTATION SYSTEM

Data to Be Collected

1. Volume of waste flow.

2. Average suspended solids concentration and variation of solids in the waste.

3. Estimation of the flotation characteristics of the wastes by the use of a laboratory flotation cell [7].

Design Procedure

1. Relate the effluent suspended solids concentration and the float solids to the calculated air/solids ratio as shown in Fig. 7.15.

2. Use a range of pressure of 30 to 60 psig (2 to 4 atm). For the desired effluent suspended solids or float solids concentration, determine the optimum air/solids ratio from the plots developed in step 1.

3. For the flotation system with a pressurized recycle, select a suitable operating pressure and use the following equation to compute the recycle flow:

$$\text{air/solids ratio} = \frac{A}{S} = \frac{1.3 s_a R (fP - 1)}{QS_a} \tag{2}$$

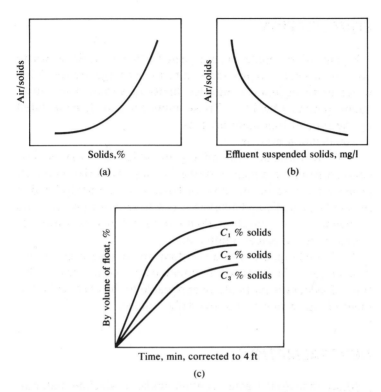

Figure 7.15 Relationship of float solids (a) and effluent suspended solids (b) to the calculated air/solids ratio. The rise rate of sludge is shown in (c).

in which s_a is the solubility of air in water at atmospheric pressure (2)(m-cm³/l), S_a the influent suspended solids (mg/l) f the fraction of saturation of air in the waste (about 0.5 for a baffled retention tank), and P the pressure (atm).

4. For a flotation system without recycle, use the following equation to compute the pressure:

$$\frac{A}{S} = \frac{1.3s_a(fP-1)}{S_a} \tag{3}$$

5. Select the overflow rate.

6. Determine the required surface area using the overflow rate obtained in step 5.

EQUALIZATION

Equalization should be provided for highly variable waste flows or strengths to provide a uniform discharge to a receiving water or to maintain controlled limits of variation to subsequent treatment facilities. These limits will usually be dictated by the type of treatment process.

Equalization may be

1. Flow through with mixing in which the volume discharged from the equalization basin varies with the inflow. A primary purpose of this type of basin is to aid neutralization (where both acidic and alkaline waste streams are present) and to equalize concentration fluctuations for neutralization or chemical or biological treatment.

2. Variable inflow with a constant rate of discharge. In this type of basin the level will fluctuate over a 24-hr period. This type of system is particularly applicable to the chemical treatment of wastewaters of a low daily volume.

NEUTRALIZATION

Many industrial wastes contain acidic or alkaline materials which require neutralization prior to discharge to receiving waters or prior to chemical or biological treatment. For biological treatment to natural waters, a pH in the biological system should be maintained between 6.5 and 8.5 to ensure optimum biological activity. The biological process itself provides a neutralization and a buffer capacity as a result of the production of CO_2, which reacts with caustic and acidic materials. The degree of preneutralization required depends, therefore, on the ratio of BOD removed and the causticity or acidity present in the waste. These requirements are discussed on page 189.

Types of Processes

1. Mixing acidic and alkaline waste streams; requires sufficient equalization capacity to effect desired neutralization. There may be a danger of toxic gases or by-products.

2. Acid wastes neutralization through limestone beds.

These may be downflow or upflow systems. The maximum hydraulic rate for downflow systems is 1 gpm/ft² (40 l/min/m²) to ensure sufficient retention time. The acid concentration should be limited to a concentration of 0.6 percent H_2SO_4 to avoid coating of limestone with nonreactive $CaSO_4$ and excessive CO_2 evolution, which limits complete neutralization. High dilution or dolomitic limestone requires longer detention periods to effect neutralization. Hydraulic loading rates can be increased with upflow beds, because the products of reaction are swept out before precipitation.

3. Mixing acid wastes with lime slurries—neutralization depends on the type of lime used. The magnesium fraction of lime is most reactive in strongly acid solutions and is useful below pH 4.2.

Neutralization with lime can be defined by a basicity factor obtained by titration of a 1-g sample with an excess of HCl, boiling 15 min, followed by backtitration with 0.5 N NaOH to phenolpthalein end point.

In lime slaking, the reaction is accelerated by heat and physical agitation. For high reactivity, the line reaction was completed in 10 min. Storage of lime slurry for a few hours before neutralization may be beneficial. Dolomitic quicklime (only the CaO portion) hydrates except at elevated temperature. Slaked quicklime is used as 8 to 15 percent lime slurry. Neutralization can also be accomplished using NaOH, Na_2CO_3, or NH_4OH.

Systems   Batch treatment is used for waste flows to 100,000 gpd (380 m³/day). Continuous treatment will employ automated pH control [where air is used for mixing, minimum air rate is 1 to 3 cfm/ft² (0.3 to 0.9 m³/min/m²) at 9 ft (3 m) liquid depth].

Basic (Alkaline) Wastes   Any strong acid can be used effectively to neutralize alkaline wastes, but cost considerations usually limit the choice to sulfuric or possibly hydrochloric acid. The reaction rates are practically instantaneous, as with strong bases.

Flue gases which can contain 14 percent CO_2, can be used for neutralization. When bubbled through the waste, the CO_2

forms carbonic acid, which then reacts with the base. The reaction rate is somewhat slower but is sufficient if the pH need not be adjusted to below 7 or 8. Another approach is to use a spray tower in which the stack gasses are passed countercurrent to the waste liquid droplets.

All the above processes usually work better with the stepwise addition of reagents, that is, a staged operation. About two stages, with possibly a third tank to even out any remaining fluctuations, are about optimum.

Control of Process The automatic control of pH for waste streams is one of the most troublesome, for the following reasons:

1. The relation between pH and concentration or reagent flow is highly nonlinear, particularly when close to neutral (pH 7.0).

2. The pH can vary over 12 decades at a rate as fast at 1 pH unit per minute.

3. The waste-stream flow rates can double in a few minutes.

4. A relatively small amount of reagent must be thoroughly mixed with a large liquid volume in a short time interval.

Advantage is usually gained by the stepwise addition of chemicals (see Fig. 7.16). In reaction tank 1, the pH may be raised to 3 to 4. Reaction tank 2 raises the pH to 5 to 6 (or any other desired end point). If the waste stream is subject to slugs or spills, a third reaction tank may be desirable to effect complete neutralization.

PROCEDURE FOR NEUTRALIZATION DESIGN

Data to Be Obtained
1. Volume of waste to be neutralized
   a. Average daily flow
   b. Variation in flow
2. Acidity or alkalinity of waste
   a. Average, after various periods of equalization
   b. Variation in acidity or alkalinity
3. pH and acidity or alkalinity of neutralized wastewaters. (This value will depend in great measure on where the neutralized effluent is to be discharged. If neutralization precedes

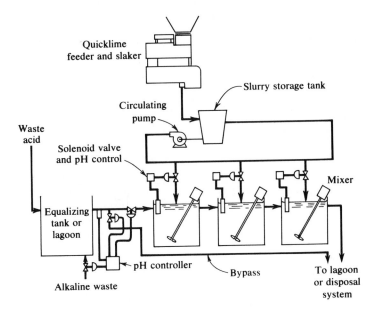

Figure 7.16 Schematic of acid-waste neutralization system.

biological treatment, the acidity or alkalinity must be reduced to a value such that the biological system can buffer the process to pH 7 to 8.)

Design Procedure Neutralization of acidic wastes is usually accomplished by the addition of slurried lime or by passing the waste through limestone beds.

1. Limestone beds
   a. A relationship is developed between flow rate, depth of limestone bed, and effluent pH (Fig. 7.17). A flow rate and a corresponding bed depth for the desired effluent pH is selected from this figure. (It should be noted that these data are based on the use of upflow units which flush out the calcium sulfate, etc., and permit escape of CO_2 generated by the neutralization reaction.)
   b. The quantity of limestone required is determined from Fig. 7.17.

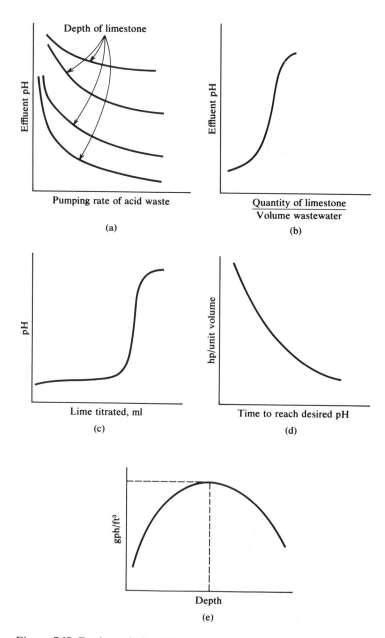

Figure 7.17 Design relationships for neutralization of acid wastewaters.

2. Lime neutralization
  a. A titration curve is developed by the addition of lime slurry to the acidic waste (Fig. 7.17).
  b. The time of reaction of the lime slurry with the waste to the desired effluent pH is related to the mixing intensity in the neutralization basin (Fig. 7.17).
  c. Depending upon the variation in influent composition, the titration-curve characteristic, and the desired effluent pH, the neutralization system can be designed as a single or multiple basin process For highly acidic wastes, a minimum of two basins is desirable, the first to raise the pH to 3.0 to 3.5 and the second to raise the pH to the desired effluent value.

3. Alkaline wastes: The basic design procedure for alkaline wastes is similar to acidic wastes. A mineral acid such as H_2SO_4 or in some cases CO_2 or scrubbed flue gas can be used.

References

1. A. Hazen, "On Sedimentation," *Trans. ASCE,* **53**, 45 (1904).
2. T. R. Camp, "Sedimentation and the Design of Settling Tanks," *Trans. ASCE,* **111**, 909 (1946).
3. W. E. Dobbins, *Advances in Sewage Treatment Design, Sanitary Engineering Division,* Metropolitan Section, Manhattan College, May 1961.
4. W. W. Eckenfelder and D. J. O'Connor, *Biological Waste Treatment,* Pergamon Press, Oxford, 1961.
5. W. W. Eckenfelder and D. L. Ford, *Laboratory and Design Procedures for Waste Treatment Process Design,* CRWR Rep. 31, University of Texas, Austin, Texas, 1969.
6. American Petroleum Institute, *Manual on Disposal of Refinery Wastes,* Vol. 1, API, New York, 1959.
7. W. W. Eckenfelder, *Industrial Water Pollution Control,* McGraw-Hill, New York, 1966.

8

Oxygen Transfer and Aeration

The oxygen-transfer process can be considered to occur in three phases. Oxygen molecules are initially transferred to the liquid surface, resulting in a saturation or equilibrium condition at the interface. This rate is very rapid. The liquid interface has a finite thickness with unique properties. The layer or film is composed of water molecules with their negative ends facing the gas phase and is estimated to be at least three molecules thick. During the second phase, the oxygen molecules must pass through this film by molecular diffusion. In the third phase, the oxygen is mixed in the body of liquid by diffusion and convection. It is assumed that at very low mixing levels (laminar flow conditions) the rate of oxygen absorption is controlled by the rate of molecular diffusion through the undisturbed liquid film (phase II). At increased turbulence levels, the surface film is disrupted and renewal of the film is responsible for transfer of oxygen to the body of the liquid[1,2,3]. This surface renewal can be considered as the frequency with which fluid with a solute concentration C_L is replacing fluid from the interface with a concentration C_S (Fig. 8.1).

The oxygen-transfer rate can be expressed

$$N = K_L A (C_S - C_L) \qquad (1)$$

where N = mass of oxygen transferred per unit time
K_L = liquid film coefficient (related to surface renewal rate; increases with turbulence or renewal rate)

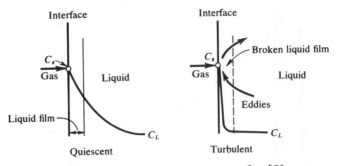

Figure 8.1 Mechanism of oxygen transfer. [7]

A = interfacial area for transfer
C_S = saturation concentration of oxygen
C_L = concentration of oxygen in the body of the liquid
Equation (1) can be reexpressed in concentration units:

$$\frac{N}{V} = \frac{dc}{dt} = K_L \frac{A}{V}(C_S - C_L) = K_L a(C_S - C_L) \qquad (2)$$

in which $K_L a$ is the overall coefficient for oxygen transfer and includes both the liquid film coefficient (K_L) and the interfacial area per unit volume (A/V). In aeration practice it is impossible to measure the interfacial area, so the overall coefficient $K_L a$ is used for aeration design. $K_L a$ can be determined from a semilogarithmic plot of ($C_S - C_L$) vs. time of aeration.

FACTORS AFFECTING OXYGEN TRANSFER

Several factors affect the performance of air aeration devices. Since aeration efficiency is estimated in water under standard conditions (zero dissolved oxygen at 20°C), corrections must be made for operation in wastewater systems. These corrections are summarized below.

Oxygen Saturation Oxygen saturation in water is related to temperature (Fig. 8.2). In wastewater, the presence of salts and other substances will affect oxygen saturation, usu-

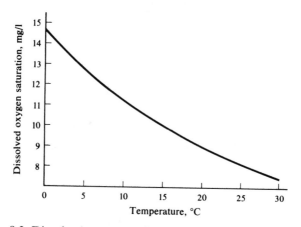

Figure 8.2 Dissolved oxygen solubility as a function of temperature. (From [4].)

ally decreasing it. Since it is not possible to estimate this value, it is necessary to measure it in the field. Because oxygen saturation is related to the partial pressure of oxygen in the gas phase (Henry's law), a correction must be made for saturation in submerged aeration devices where the partial pressure at the point of air discharge exceeds atmospheric pressure. It is convenient to use the average saturation at the tank middepth. This can be estimated from

$$C_{S_M} = C_S\left(\frac{P_b}{29.4} + \frac{O_t}{42}\right) \tag{3}$$

where C_{S_M} = oxygen saturation at the aeration tank middepth
P_b = pressure at tank bottom, psig
O_t = oxygen in the exit gas, percent

Temperature The oxygen-transfer coefficient $K_L a$ will increase with increasing temperature. The relationship most commonly used is

$$K_{La(t)} = K_{La(20°C)} \cdot 1.02^{(T-20)} \tag{4}$$

where T is the temperature in °C. Recently, Landburg et al.[5] found a temperature coefficient of 1.012 for surface aeration units.

Dissolved Oxygen Level For biological-treatment systems removing carbonaceous organics, the operating dissolved oxygen level, C_L, will vary between 0.5 and 1.5 mg/l. When nitrification is to be achieved, the dissolved oxygen level will be in excess of 2.0 mg/l.

Effect of Waste Characteristics on Oxygen Transfer

The presence of surface-active agents will have a marked effect on the oxygen-transfer rate as they affect both the liquid film coefficient K_L and the A/V ratio and hence K_La. This effect will be reflected by changes in concentration of surfactant and by changes in the turbulence and mixing in the system.

A surfactant will concentrate at an interface such that the interfacial concentration will be greater than that in the body of the liquid. As a result, a "film" of adsorbed surfactant molecules is concentrated at the interface, which provides a barrier to diffusion of oxygen across the interface.

The changes in transfer rate in the presence of surface-active materials is defined by the coefficient, α, which is $K_La_{(waste)}/K_La_{(water)}$. Increasing the concentration of surfactant will decrease α until the interfacial surface is saturated. Further increases will not affect α (Fig. 8.3a). In bubble aeration, the presence of surfactants will markedly decrease the bubble size and hence increase A/V. Under these conditions, it is possible for α to increase in some cases to values greater than 1.0 (Fig. 8.3b), because the increased effect of A/V exceeds the decrease in K_L caused by the surface barrier.

The degree of turbulence in the system will also affect α.

Under laminar conditions (approaching a stagnant film surface) there is substantially no effect on α because the resistance in the bulk of solution to oxygen transport exceeds the combined interfacial resistance[6]. This condition would rarely be encountered in aeration practice. Under moderately turbulent conditions, a maximum depression occurs because the interfacial resistance to molecular diffusion by the adsorbed surfactant molecules controls the transfer rate.

At high degrees of turbulence, α approaches unity as a result of the high surface-renewal rates, resulting in an inability to establish an adsorption equilibrium at the interface (Fig. 8.4). α often exceeds unity as a result of the increased A/V values as-

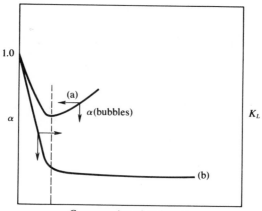

Figure 8.3 Effect of concentration of surfactant on oxygen transfer.

sociated with turbulent conditions at the surface and bubble entrainment (Fig. 8.4).

The coefficient α can be expected to increase or decrease

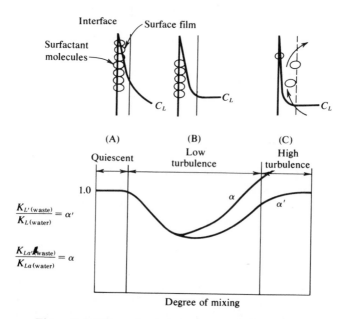

Figure 8.4 Effect of turbulence on oxygen transfer. [7]

and approach unity during the course of biooxidation because the substances affecting the transfer rate are being removed in biological process. In order to design the aeration system, the standard transfer efficiency (STE) is corrected for the waste conditions:

$$\text{oxygen transfer in waste} = \text{STE} \cdot \frac{C_{SW} - C_L}{9.1} \cdot \alpha \cdot 1.02^{(T - 20)}$$

AERATION EQUIPMENT

The aeration equipment commonly employed in the waste-water-treatment field consists of air diffusion units, turbine aeration systems in which air is released below the rotating blades of a submerged impeller, and surface aeration units in which oxygen transfer is accomplished by high surface turbulence and liquid spray.

Diffused Aeration Equipment Diffused aeration equipment is basically of two types: that which produces a small bubble from a porous media, and that which uses a large orifice or a hydraulic shear device to produce larger air bubbles.

Porous media are either tubes or plates constructed of carborundum or tightly wrapped plastic wrap or nylon. Tubes are placed at the side wall of the aeration tank perpendicular to the wall and generate a rolling motion to maintain mixing. Minimum and maximum spacings are required to maintain solids in suspension and to avoid bubble coalescence, respectively. The size of bubbles released from this type of diffuser range from 2.0 to 2.5 mm.

To maintain adequate mixing, the maximum width of aeration tank is approximately twice the depth. This width can be doubled by placing a line of diffusion units along the center line of the aeration tank.

Large-bubble air-diffusion units will not yield the oxygen-transfer efficiency of fine-bubble diffusers, because the interfacial area for transfer is considerably less. These units have the advantage, however, of not requiring air filters and of gen-

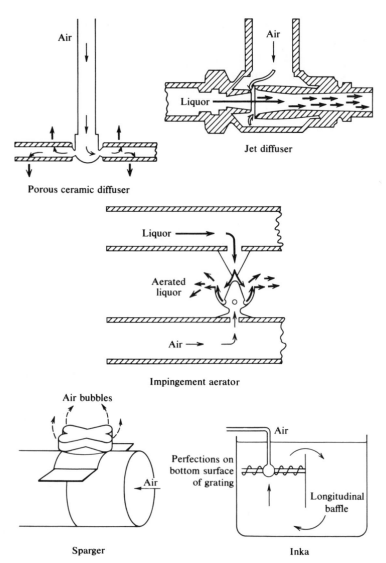

Figure 8.5 Schematic representation of aeration devices. (From Eckenfelder and O'Connor [8].)

erally requiring less maintenance. Large-bubble diffusers are placed along the side wall of an aeration tank in a manner similar to porous diffusers. These units generally operate over a wider range of air flows per unit. Commerical units include the sparger, the hydraulic shear diffuser, the Venturi diffuser, and the Inka diffuser (as shown in Fig. 8.5).

The variables affecting the performance of diffused aeration units are air flow rate, tank liquid depth, and tank width. These are shown for a sparger unit in Fig. 8.6.

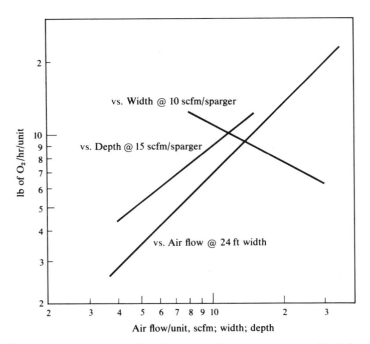

Figure 8.6 Oxygen transfer characteristics of a sparger unit. (After Eckenfelder and Ford [7].)

The performance of all air diffusion units can be expressed by the relationship

$$N = C G_s^{(1-n)} \frac{H^m}{W^p} (C_{SW} - C_L) \cdot 1.02^{(T-20)} \cdot \alpha \qquad (5)$$

where N = lb of O_2/hr/aeration unit

C = constant for the aeration unit

G_s = air flow, scfm/aeration unit

H = liquid depth, ft
W = aeration tank width, ft
C_L = dissolved oxygen concentration in liquid
T = temperature, °C
α = oxygen-transfer coefficient of the waste
n,m,p = exponents characteristic of the aeration device

Aeration performance characteristics are summarized in Table 8.1. An air-diffusion unit is shown in Fig. 8.7.

Table 8.1. Diffused Aeration-Performance Characteristics

Unit	C	n	Conditions
Seran tubes	0.16	0.90	9-in. spacing, wide band; 14.4-ft depth, 24-ft width
Seran tubes	0.17	0.81	9-in. spacing, narrow band; 14.4-ft depth, 24-ft width
Seran tubes	0.15	0.92	9-in. spacing, narrow band; 14.4-ft depth, 24-ft width
Spargers	0.081	1.02	24-in. spacing, wide band; 14.8-ft depth, 24-ft width
Spargers	0.062	1.02	9-in. spacing, narrow band; 14.8-ft depth, 24-ft width
Spargers	0.064	1.02	9/32-in orifice; 25-ft width 15-ft depth
Spargers	0.068	1.02	13/64-in. orifice; 25-ft width 15-ft depth
Plate tubes	0.35	0.49	Single row; 25-ft width, 15-ft depth
Plate tubes	0.20	0.80	Double row; 25-ft width, 15-ft depth
INKA system	0.036	0.95	6.8-ft width, 6-ft depth, 2.6-ft. submergence

Turbine Aeration   In turbine aeration, air is discharged from a pipe or sparge ring beneath the rotating blades of an impeller. The air is broken into bubbles and dispersed throughout the tank contents. Present commercial units employ one or more submerged impellers and may employ an additional impeller near the liquid surface for oxygenation from induced

Figure 8.7 Air-diffusion system. (Courtesy of Chicago Pump Company.)

surface aeration. A typical turbine unit is shown in Fig. 8.8.

In addition to air flow, both the diameter and the speed of the impeller will affect the bubble size and velocity, thus influencing the overall transfer coefficient, $K_L a$.

$$N = C R^x G_s^n d^y (C_s - C_L) \qquad (6)$$

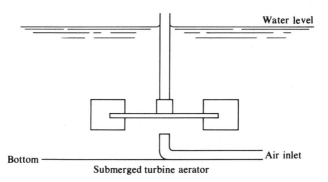

Figure 8.8 Turbine aeration unit. (Courtesy of Mixing Equipment Co.)

where R = impeller peripheral speed, fps
   d = impeller diameter, ft
   x,y = exponents characteristic of the aeration device

The power drawn by the turbine aerator can be computed from the relationship

$$\text{horsepower} = Cd^n R^m \tag{7}$$

When air is applied beneath the impeller, the actual power drawn is reduced because a less dense mixture is being pumped by the turbine. This relationship has been shown by Eckenfelder and O'Connor[8] and others.

Quirk[9] developed two significant correlations between the oxygen-transfer efficiency and the power supplied to the system from the rotor (HP_R) and the air flow (G_S), as shown in Fig. 8.9.

In most cases, P_d^* occurs near 1.0. (This implies an equal power expenditure by the turbine and the blower.) At extremely high air rates, ($P_d < 1.0$), large bubbles and flooding of the impeller yield poor oxygenation efficiencies, while at very low rates ($P_d > 1.0$) too much turbine horsepower is being expended in fluid mixing.

Variation in oxygen input to the system can most easily be adjusted by varying the air rate under the impeller. This in turn will change P_d. The anticipated range of operation should

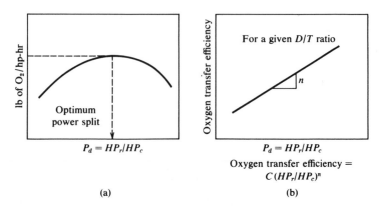

$$P_d = HP_r / HP_c$$

Oxygen transfer efficiency = $C(HP_r/HP_c)^n$

(a)   (b)

Figure 8.9 Design relationships for turbine aeration.

cover the maximum range of operation efficiency as related to P_d.

Available data indicate that the oxygenation efficiency of turbine aerators in water should vary from 2.5 to 3.0 lb of O_2/hp-hr (1.52 to 1.82 kg of O_2/kwh and reducer losses).

Surface Aerators During the past 5 years, surface aerators have found increasing application in activated sludge plants and aerated lagoons in the United States. This is largely due to improved design, resulting in increased oxygen-transfer capacity.

Several types of surface aerators are in use today. The brush aerator employs a rotating steel brush which sprays liquid from the rotating blades. Mixing is accomplished by an induced velocity below the rotating element. Oxygenation capacity is expressed per foot of rotor and is related to the submergence of the blades.

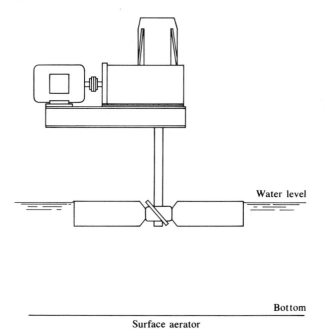

Surface aerator

Figure 8.10 Surface aeration unit. (Courtesy of Mixing Equipment Co.)

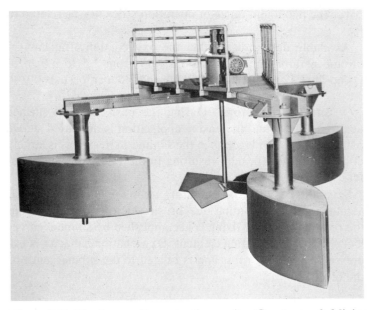

Figure 8.11 Floating surface aeration unit. (Courtesy of Mixing
Equipment Co.)

Bladed-surface aerators pump liquid from beneath the
blades and spray the liquid across the surface of the water.
Some units employ a draft tube. A recently developed unit employs a submersible pump which delivers liquid against a plate
from which it is sprayed across the water surface. These units
are float-mounted on polyethylene "doughnuts." Several surface aerators are shown in Fig. 8.10, 8.11, and 8.12.

Oxygen transfer in most types of surface aerators may be
considered to occur in two ways: transfer to droplets and thin
sheets of liquid sprayed from the blades of the unit, and transfer at the turbulent liquid surface and from entrained air bubbles where the spray strikes the surface of the liquid.

Since the total transfer is primarily related to the total surface generated and the quantity of liquid pumped, performance
will be related to the submergence of the impeller in the liquid
and the speed and diameter of the rotating element. For most
units, the lb of oxygen/hp-hr transferred remains substantially
constant over a wide range of unit sizes at optimum submer-

Figure 8.12 Floating surface aeration unit. (Courtesy of Wells Products, Inc.)

gence. This equipment can therefore be designed by adjusting the oxygenation capacity for anticipated operating conditions. It should be noted, however, that this relationship does not include geometrical or power-level considerations.

Most surface aeration units will transfer 3.2 to 3.8 lb of O_2/hp-hr (1.94 to 2.30 kg of O_2/kwh) depending on tank geometry, surface area, and unit construction.

It should be recognized that aerators serve two functions in the biological process: the transfer of the required oxygen and including sufficient mixing to maintain uniform oxygen throughout the basin, as in the case of the aerobic-facultative lagoon, and keeping the biological solids in suspension in the aerobic lagoons and the activated sludge process. For conventional or high-rate organic loadings, the power required for oxygen transfer is usually considerably in excess of that required for mixing. In large aerobic-facultative lagoons or extended aeration systems, however, power for mixing may control the aerator design. Figure 8.13 shows the power relation-

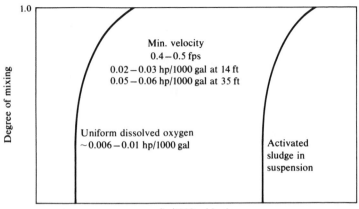

Figure 8.13 Power requirements for mixing in biological treatment
            processes.

ships for dispersion of oxygen and for maintenance of activated sludge in suspension. Experiments on one type of surface aerator indicated a minimum power level for oxygen dispersion of 6 to 8 hp/mg [1.18 to 1.58 kw/m³(10³)] of basin volume. (This power level is defined as that which will maintain a
dissolved oxygen concentration ±0.4 mg/l throughout the
basin.) Experiments in the United States and Germany[10]
have shown that a velocity of 0.4 to 0.5 fps (12.2 to 15.2 cm/
sec) may be necessary to maintain normal activated sludge in
suspension. Sludges with a high inorganic content may require
higher velocities.

The maintenance of solids in suspension limits the depth of
aeration basin with some types of aeration units. Data have indicated that units without a draft tube have a maximum depth
of 12 ft (3.65/m) (unless supplementary mixing is provided)
and units with a draft tube to 15 to 17 ft (4.65 to 5.4/m).

It has recently been shown that the oxygenation of most surface aeration units varies with the volume of liquid under aeration. This is logical when one considers that the amount of oxygen transferred depends upon the interfacial surface exposed
and the total volume of liquid pumped. When the liquid strikes
the tank surface, turbulent mixing and air entrainment will result in additional oxygen transfer. If the aeration liquid volume

for a given size aerator is increased, energy dissipation throughout the larger volume will result in less turbulence and entrainment and hence a reduced oxygen transfer. The respective portions of the oxygen transferred in the liquid spray and the associated turbulence and entrainment will depend on the power level. In an infinite volume, for example, virtually all of the oxygen transfer would result from the liquid spray alone. As a result of these considerations the oxygenation efficiency can be related to liquid volume:

$$N_o \text{ (lb of O}_2/\text{hp-hr} = KP_v + N_s$$

where N_o = total oxygen transferred under standard conditions per unit horsepower

P_v = horsepower per 1000 gal or watts/m³ of basin liquid

N_s = oxygen transferred from the liquid spray

K = constant characteristic of the aeration device

Aeration-performance data in water for a surface aerator are shown in Fig. 8.14.

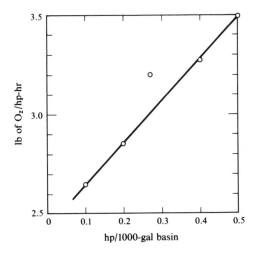

Figure 8.14 Effect of basin volume on oxygenation efficiency.

PROCEDURE FOR THE DESIGN OF AERATION SYSTEMS

Data to Be Collected
1. Volume of aeration tank (from biological oxidation requirements).
2. Oxygen requirements (see biological treatment design).
3. Type of aeration system.
4. Operating temperature.
5. Oxygen-transfer coefficient, α.

Design Procedure
1. Turbine aeration
   a. Select a D/T ratio (diameter of turbine/diameter of aeration tank) between 0.1 and 0.2.
   b. From the equipment manufacturer's data, determine the optimum power split (hp of rotor/hp of compressor), oxygenation capacity, and the transfer efficiency. (See Fig. 8.7.)
   c. Select a liquid depth between 15 and 20 ft (4.55 and 6.1/m) (in some cases deeper liquid depths may be employed). The aeration tank should be either a square or a rectangle.
   d. Determine the number of turbine units based on the approximate basis of one for every 900 to 2500 ft^2 (83.5 to 2.32 m^2).
   e. Determine the air-flow rate to supply the required oxygen from the optimum transfer efficiency (Fig. 8.7).
   f. Determine the operating compressor horsepower using the relationship hp $= [G_s(\text{psi})(144)/(33,000)E]$, in which G_s is the air flow rate and E the motor and blower efficiency, and correct it for waste conditions.
   g. Determine the turbine hp from the power split (Fig. 8.7).
   h. Calculate the aeration efficiency, correcting for waste conditions.
2. Diffused aeration
   a. Select a tank depth between 10 and 16 ft (3.5 and 4.9 m).

b. Plan the size of the aeration tank so that its maximum width is approximately twice the tank depth. This is necessary in maintaining adequate mixing.

c. Select the air-flow rate for the air-diffusion unit.

d. Compute C_{SM} (oxygen concentrations at tank mid-depth) using equation (3).

e. The oxygenation capacity of the aeration unit (N) is computed using equation (5).

f. Determine the number of aeration units required to transfer the required amount of oxygen.

g. Determine the spacing of the aeration units from the relationship

$$\text{spacing} = \frac{\text{length of tank}}{\text{No. of aeration units}}$$

The minimum and maximum spacings should be 6 in. (15.2 cm) and 24 in. (61 cm) to 30 in. (76), respectively, to maintain solids in suspension and to minimize bubble coalescence. If the spacings calculated fall outside this range, double rows or an adjustment in the number of units should be made.

h. Compute the total air flow and the required horsepower of the blower using the relationship

$$\text{hp} = \frac{(\text{psi})(\text{scfm})(144)}{(33,000)\,E}$$

Usually a 6- to 10-psi pressure is adopted on the blower and the efficiency depends on the type of the diffused aeration unit.

i. Compute the oxygenation efficiency.

3. Surface aeration

a. Select a value of C_L depending on the process involved (that is, BOD removal only or nitrification).

b. Estimate α (oxygen-transfer coefficient for the waste) from scale studies.

c. Compute C_{SW} (oxygen saturation concentration for waste).

d. From the performance characteristics curve of the sur-

face aerator, determine the value hp/1000 gal for a selected N_o (Fig. 8.11).

e. Compute N (oxygen-transfer capacity of the aeration
unit for field conditions) using the relationship

$$N = \frac{N_o(C_{SW} - C_L)}{9.1} \cdot \alpha \cdot 1.02^{(T - 20)}$$

f. Compute the total hp required to transfer the desired
amount of oxygen.

g. Recompute from the hp calculated, the power level as
hp/1000 gal and check to see if the power level is sufficient for mixing and keeping the solids in suspension.
Power levels of approximately 0.008 and 0.09 hp/1000
gal (0.00156 to 0.0176 kw/m³) are required to transfer
the oxygen by diffusion and to keep the solids in suspension, respectively.

References

1. W. K. Lewis and W. G. Whitman, "Principles of Gas
Absorption," *Ind. Eng. Chem.*, **16**, 1215 (1924).

2. P. V. Danckwertz, "Significance of Liquid Film Coefficients in Gas Absorption," *Ind. Eng. Chem.*, **43**, 6 (1951).

3. W. E. Dobbins, "Mechanism of Gas Absorption by Turbulent Liquids," in *Advances in Water Pollution Research*,
Vol. 2, Pergamon Press, Oxford, 1964, p. 61.

4. "Solubility of Atmospheric Oxygen in Water," *J. Sanit.
Eng. Div., ASCE*, No. SA4, 41 (1960).

5. G. Landberg et al., "Experimental Problems Associated
with the Testing of Surface Aeration Equipment," *Water
Research* (in press).

6. K. H. Mancy and D. Okun, "The Effects of Surface Active Agents on Aeration," *J. Water Pollution Control
Federation,* **37** (1965).

7. W. W. Eckenfelder and D. L. Ford, *Advances in Water
Quality Improvement,* Vol. I, University of Texas Press,
Austin, Texas, 1967.

8. W. W. Eckenfelder and D. J. O'Connor, *Biological
Waste Treatment*, Pergamon Press, Oxford, 1961.

9. T. P. Quirk, *Optimization of Gas-Liquid Contacting Systems*, unpublished report, 1962.
10. E. H. E. Knop, *Versuche mit Verschiedenen Beluftungssystemen im technischen Massstab*, Emschergenossenschaft, Essen, West Germany, 1964.

9
Biological Treatment

Many types of microorganisms are active in the breakdown of organic matter and the resulting stabilization of organic wastes. Most biological systems treating organic wastes depend upon heterotropic organisms which utilize organic carbon as an energy source and as a carbon source for cell synthesis. Anaerobic heterotrophs obtain their energy from organic compounds in the absence of molecular oxygen.

Autotropic organisms, on the other hand, do not require an organic carbon source. Chemosynthetic autotrophs obtain energy from the oxidation of inorganic compounds such as nitrogen or sulfur. Photosynthetic autotrophs utilize solar energy for the synthesis of carbon dioxide to cellular protoplasm and produce molecular oxygen as a by-product.

CELLULAR GROWTH IN BIOLOGICAL SYSTEMS

It has been shown by McCarty [1] that the similarity in the biochemistry of synthesis of all microorganisms under a wide variation in environment permits calculation of cellular yields from thermodynamic considerations.

In the stabilization of an organic substrate, a portion of the energy obtained from the reaction is used for biological synthesis and the remainder to satisfy the energy requirements for growth. A small portion of the energy is used for cellular maintenance (Fig. 9.1).

In aerobic growth, energy is released from the conversion of organic carbon, resulting in considerable energy being available for synthesis and hence a relatively high yield coefficient, *a*. McCarty [1] obtained values for *a* varying from 0.30 to 0.51

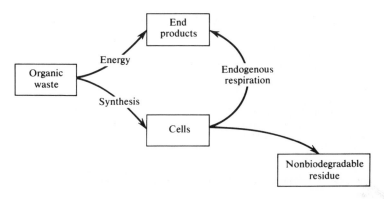

Figure 9.1 The Mechanism of aerobic biological oxidation.

for glucose, aniline, lactate, and acetate. Servizi and Bogan [2] showed a value of 0.39 for a variety of compounds. Eckenfelder and O'Connor [3] showed yield coefficients varying from 0.37 to 0.46 for several readily degradable organic wastes.

In anaerobic systems, less energy is obtained from the organic conversion, and hence the growth yield is much less than for aerobic systems. Yield coefficients varying from 0.032 to 0.27 were found depending on the substrate. A detailed study by Andrews et al. [4] on a synthetic soluble substrate showed a yield coefficient of 0.14. It should be realized that most of these growth yields include both the acid formers and the methane organisms.

In autotropic growth, considerable energy must be expended to convert CO_2 to an intermediate for cell synthesis. As a result, yield coefficients of less than 0.1 are usually obtained. Several sources reported a values for *Nitrosomonas* of 0.015 to 0.03 and 0.02 for *Nitrobacter*.

The measured yield coefficients for various systems are summarized in Table 9.1 and include the influence of volatile suspended solids originally present in the waste and hence denoted as a.

In practice, the presence of other organic suspended solids will increase the apparent yield coefficient, as graphically determined in Fig. 9.2. Sludge-accumulation data from various sources for domestic sewage are shown in Fig. 9.3.

Table 9.1.   Aerobic Biological Waste-Treatment Parameters[a]

Waste	a (BOD$_5$ basis)	a' (BOD$_5$ basis)	$b,$ l/day	k
Domestic	0.73	0.52	0.075	0.017–0.03
Refinery	0.49–0.62	0.40–0.77	0.10–0.16	0.074
Chemical and				
petrochemical	0.31–0.72	0.31–0.76	0.05–0.18	.0029–0.018
Brewery	0.56	0.48	0.10	—
Pharmaceutical	0.72–0.77	0.46	—	0.018
Kraft pulping				
and bleaching	0.5	0.65–0.8	0.08	—

[a] All parameters include the effect of influent suspended solids.

The sludge yield for a biological system can be estimated from the relationship

$$\Delta X_v = S_0 + as_r - bX_v \tag{1}$$

or

$$\Delta X_v = S_0 + as_r - bX_d \tag{1a}$$

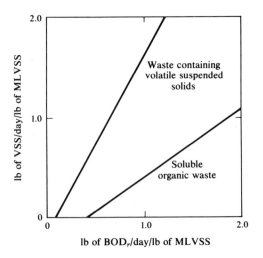

Figure 9.2 Sludge-production characteristics for a soluble waste and a waste containing volatile suspended solids.

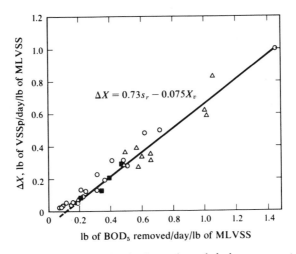

Figure 9.3 Sludge production in the activated sludge process treating domestic sewage. (After Wuhrmann [15])

where X_d = biodegradable mixed-liquor suspended solids
X_v = mixed-liquor volatile suspended solids
S_0 = influent volatile suspended solids not degraded
s_r = BOD removed

Equation (1) is only approximately correct for conventional-type systems because the term X_v includes nonbiodegradable residues. Equation (1a) is more correct because the term X_d is only the biodegradable mass of biological solids in the system. Equation (1a) should be used when considering the extended aeration process. It should be emphasized that in many cases it is not possible to divide the sludge yield ΔX_v into the contribution by microbial synthesis and the accumulation of volatile suspended solids originally present in the waste. The experimental coefficient a is then used for engineering design (Table 9.1 and Figs. 9.2 and 9.3), and equation (1) becomes

$$X_v = \overline{a}s_r - bX_v \qquad (1b)$$

It should be noted that for a soluble waste, $\bar{a}$ is approximately equal to a.

In a flowthrough system, without recycle, the concentration of solids in the effluent will be equal to the concentration in the reactor, and the sludge yield is equal to the solids lost in the effluent. Equation (1) can be reexpressed for the flowthrough system as

$$X_v Q = S_0 Q + a s_r Q - b X_v V$$

Dividing by Q and rearranging,

$$X_v = \frac{S_0 + a s_r}{1 + bt} \tag{2}$$

in which X_v is the concentration of volatile suspended solids maintained in the reactor.

ENDOGENOUS METABOLISM

Endogenous metabolism occurs in all cells in which energy is utilized for cellular maintenance. Endogenous metabolism may be defined by the coefficient, b, which has the units of reciprocal time; that is, there is a fractional decrease in cell mass per day. It should be noted that the coefficient, b, applies to the degradable cellular solids, particularly in the case of total oxidation-type systems [3]. Endogenous values reported for several systems are summarized in Table 9.1.

When high aeration solids are carried, the accumulation of nonbiodegradable mass reduces the percentage of active organisms present in the system. For example, the data of Wuhrmann [5] would indicate 59 percent active organisms at 6000 mg/l relative to substantially total activity at less than 500 mg/l of aeration solids.

OXYGEN UTILIZATION IN AEROBIC SYSTEMS

In aerobic systems, the portion of substrate not utilized for cellular synthesis uses oxygen for energy. In addition, oxygen is used for cellular maintenance (endogenous respiration). The resulting relationships are

$$O_2 = a's_r + b'X_v \tag{3}$$

and

$$O_2 = a's_r + b'X_d \tag{3a}$$

When volatile suspended solids are undergoing slow degradation in the aeration system, a' will reflect this oxygen usage. Also, nitrification will vastly increase the value of a'. As in the case of the synthesis relationships, equation (3) can generally be used for conventional systems, while equation (3a) should be used for extended aeration or total oxidation systems (see Fig. 9.4).

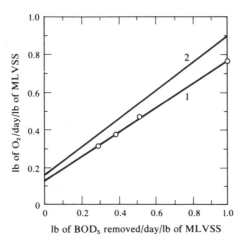

Figure 9.4 Oxygen requirements in biological oxidation processes. Curve 1, sewage and textile waste; curve 2, petrochemical waste.

When a material balance is based on COD or BOD_u, $a + a' = 1$, because the organic substrate carbon results in the production of CO_2 or the formation of biological cells. When based on BOD_5, the following conversion must be made:

$$a_5' = \frac{1}{BOD_5/BOD_u} - 1.42a_5 \qquad (4)$$

KINETICS OF REMOVAL OF ORGANIC SUBSTRATES

Several mathematical models have been offered to explain the mechanism of BOD removal by biological treatment processes [6,7]. All of these models have shown that at high BOD levels, the rate of BOD removal per unit of cells will remain constant to a limiting BOD concentration below which the rate will become concentration dependent and decrease. The rate of cell growth may continue at a maximum longer than the rate of BOD removal due to assimilation of stored BOD.

Wuhrmann [8] has shown that single substances are removed by a zero-order reaction to very low substrate levels. When a mixture of substances exists, all of which are being removed at different rates, a constant maximum removal rate will exist until the most rapidly removed of the substances is completely removed. The overall rate will then decrease. As other substances are progressively removed, the overall rate decreases further. The overall removal rate may approximate first- or second-order kinetics, depending on the substrate mixture [9]. As Gaudy et al. [10] have shown, sequential substrate removal may also yield a decreasing overall rate. The removal of a mixture of organics is shown in Fig. 9.5.

The Michaelis–Menton relationship has also been employed to define the microbial growth rate:

$$\frac{ds}{dt} = \mu_m \frac{s}{K_s + s} \cdot X \qquad (5)$$

and

$$\frac{dX}{dt} = a \frac{ds}{dt} - bX \qquad (6)$$

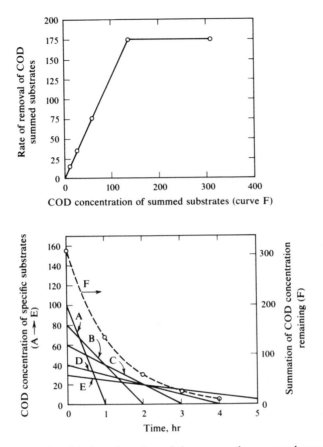

Figure 9.5 Graphical explanation of the zero-order removal concept.

Combining equations (5) and (6) yields

$$\frac{dX/dt}{X} = \frac{a\mu_m s}{K_s + s} - b \qquad (7)$$

The removal rate per unit of organisms, $(ds/dt)/X$, is equal to $\mu_m s/(K_s + s)$. As $(dX/dt)/X$ is equal to 1/sludge age $(1/G)$ or in the case of a flowthrough system $1/t$, therefore

$$\frac{1}{t} = \frac{1}{G} = \frac{a\mu_m s}{K_s + s} - b \qquad (7a)$$

At high concentrations of s when substrate concentration is not limiting microbial growth, $s > K_s$ and equation (5) reduces to

$$\frac{ds}{dt} = \frac{\mu_m}{a} \cdot X \tag{8}$$

When the increase in biological solids, ΔX, is less than $2X_a$, equation (8) can be expressed

$$\frac{as_r}{X_V t} = \mu_m \tag{8a}$$

At very low concentrations of s, $K_s < s$ and equation (5) reduces to

$$\frac{ds}{dt} = \frac{\mu_m}{K_s} sX \tag{9}$$

For most individual substrates, in batch or plug-flow conditions, K_s is very low, and equation (8) applies to very low substrate levels. Waste mixtures such as those encountered in sewage or industrial wastes contain many substrates, yielding varying growth rates. The combined effect is to produce an average rate, k_a. The relationship can usually be approximated by a pseudo-first-order or a pseudo-second-order reaction:

$$\frac{s}{s_0} = e^{-k_a x_v t} \tag{10}$$

or

$$\frac{s}{s_0} = \frac{1}{1 + k_a X_v t} \quad \text{or} \quad \frac{s_0 - s}{X_v t} = k_a s \tag{11}$$

k_a is usually referred to as the average removal-rate coefficient, k.

Most of the data obtained from the treatment of wastes with low initial BOD values (500 mg/l) can be correlated according to equations (10) or (11).

COMPLETELY MIXED SYSTEMS

In a completely mixed system, the effluent concentration is approximately equal to the concentration of soluble substrate remaining in the aeration tank, and a material balance can be developed:

$$\frac{Qs_a - Qs_e}{X_v} = -\frac{ds}{dt}V$$

The removal rate, after dividing by Q, becomes

$$\frac{s_a - s_e}{X_v} = -\frac{ds}{dt} \tag{12}$$

For a first-order-type reaction, equation (12) becomes

$$\frac{s_a - s_e}{X_v t} = \frac{s_r}{X_v t} = k_a s \tag{12a}$$

The data from most completely mixed systems in which the effluent concentration of soluble BOD or COD is low can be correlated according to equation (12). Several examples are shown in Fig. 9.6.

SLUDGE AGE

Sludge age is defined as the average length of time the microorganisms are under aeration. In a flowthrough system, sludge age is the dilution rate ($1/Q$). For growth to occur and to effect BOD removal, the growth rate becomes

$$\mu = \frac{1}{G} = R_h \tag{13}$$

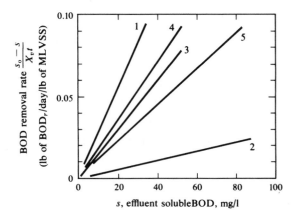

Figure 9.6 BOD-removal characteristics from various organic wastes. Curve 1, synthetic rubber; curve 2, petrochemical curve 3, rendering; curve 4, sewage—Kassel; curve 5, sewage—Flensburg.

in which G is the sludge age and R_h the dilution rate.

In a system with sludge recycle and a wastage of excess sludge, sludge age is defined as

$$G = \frac{X_v}{\Delta X_v} = \frac{X_v}{as_r - bX_v} \tag{14}$$

where X_v and ΔX_v are expressed in lb (kg) and lb/day (kg/day), respectively.

Equation (13) also applies relative to the limiting growth rate for specific organisms.

It should also be noted that small residual BOD or COD, s, will remain even after long periods of aeration, because auto-oxidation of the sludge results in resolubilization of cellular material which is subsequently used for synthesis.

It should be recognized that in any one plant as the composition of the waste changes, the overall removal rate may also change, owing to variation in removal rate of particular constituents and their initial concentration. The reaction rates observed for several aerobic systems in accordance with equation (12a) are summarized in Table 9.1.

SLUDGE SETTLING

Sludge settling and compaction characteristics are a primary requisite to successful operation of the activated sludge process. With a poor settling sludge, solids carryover in the effluent will contribute to the BOD (due to endogenous respiration of the activated sludge solids in the BOD bottle). The poor compaction will result in a low concentration of return sludge solids, which in turn will limit the mixed-liquor suspended-solids level.

A poor settling or bulking sludge can be the result of the propagation of filamentous organisms (i.e. *Sphaerotilus*) or from diffuse bacterial growth.

Many filamentous growths are obligatory aerobes which flourish in the presence of an available carbon source such as glucose. At low concentrations of dissolved oxygen in the mixed liquor (< 0.5 mg/l) there is little oxygen penetration into the biological floc and only a small fraction of the bacterial mass will exhibit aerobic growth [5,11]. The filaments, on the other hand, have a very high surface area/volume ratio and will quickly outgrow the bacterial population [12]. High oxygen tension will favor the growth of floc-forming bacteria. When the carbon source is exhausted, the filaments will tend to disappear from the system.

Many filamentous organisms are aerobic and can be destroyed by prolonged periods of anaerobiasis. Most of the bacteria, on the other hand, are facultative and can exist for extended periods without oxygen. Although available data are somewhat contradictory, it would appear that at least 6 hr under anaerobic conditions are necessary to eliminate the growth of these filaments.

In practice, culture control can frequently be achieved by holding the mixed liquor in the final settling tank for the period sufficient to eliminate the filaments.

Endogenous sludges can be maintained under anaerobic conditions for long periods because the exogenous food supply has been exhausted. At high loadings, however, gasification due to anaerobic activity may occur, resulting in rising sludge. At very low loadings, denitrification may also cause rising sludge. This is shown in Fig. 9.7.

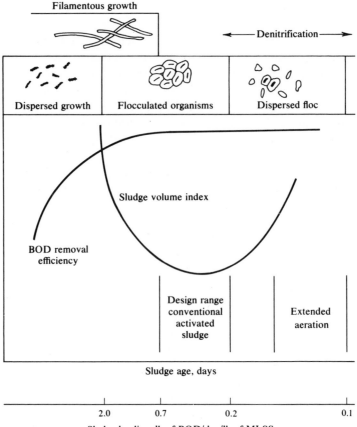

Figure 9.7 Flocculation and settling characteristics of activated sludge as related to organic loading.

Various investigators have related sludge bulking to organic loading. Indications have shown that bulking becomes progressively more severe as the loading exceeds 0.5 lb of BOD/day/lb, MLSS (0.5 kg of BOD/day/kg of MLSS)[14]. Von der Emde[16], however, reported excellent settling characteristics with loadings in excess of 2.0 lb of BOD/day/lb of MLSS. This apparent contradiction can possibly be explained by considering the cause of bulking and its relationship to the operation of the process.

Sludge bulking can be related to the growth rate or metabolic activity of the sludge, which in turn is related to the food/microorganisms ratio, F/M. At very high F/M ratios, the organisms have a maximum growth rate (log growth phase) and flocculation does not occur [15], and filamentous organisms can develop. At very low loadings, unoxidized fragments of floc remain in suspension, resulting in poorer settling. These regimes are schematically shown in Fig. 9.7. As the F/M is decreased, the organisms flocculate.

Since the significant parameter is the concentration of food in contact with the organisms, the geometry of the system and the mode of introduction of the waste deserve consideration. For example, in a long, rectangular aeration tank or in a batch-treatment system, the sludge is initially contacted with the sewage entering the system and the F/M ratio is high at the head end of the tank (or the start of aeration in the case of a batch-treatment system). A diffuse floc developed under these conditions could persist throughout the aeration period. By contrast, when a complete mixing system is used, and the sewage distributed throughout the aeration tank, the F/M ratio at anytime is low (i.e. the sludge is always in contact with a BOD concentration approximately equal to the effluent). As a result, high loading levels expressed as lb of BOD/day/lb of MLSS, can still yield a low F/M ratio and a dense floc.

When considering the effect of the F/M ratio on metabolic activity, the availability of the substrate must be considered. Soluble sugars, for example, are immediately available to the organism and consequently yield an immediate growth response. Suspended organics, on the other hand, must undergo sequential breakdown to simpler substrates before being available for synthesis. The growth response is therefore much slower, even at high F/M ratios.

In summary, floc characteristics as related to growth rate will be influenced both by the availability of the substrate and the mode of introduction of the waste to the system. Typical results for an industrial waste are shown in Fig. 9.8.

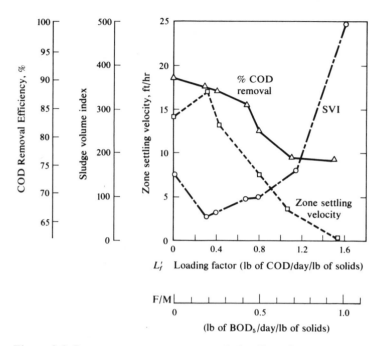

Figure 9.8 Parameter response to organic loading—brewery waste. (After Ford [13].)

EFFECT OF TEMPERATURE

One of the significant variables in the selection of type of process is the effect of temperature on process performance. The temperature effect on the reaction rate can be expressed by the relationship

$$K_T = K_{20°C} \Theta^{(T - 20)} \tag{15}$$

The coefficient Θ for microbial activity expressed as respiration rate has been reported as 1.074 and 1.085 by Wuhrmann [8] and Eckenfelder and O'Connor [3], respectively. It has been shown, however, that this coefficient varies markedly with the type of process (activated sludge, aerated lagoons, etc.). The reported coefficients are summarized in Table 9.2.

Table 9.2. Temperature Coefficients for Microbial Activity

Process	Θ	Reference
Activated sludge		
Loadings < 0.5 kg of BOD/day/kg of MLSS	1.0	[15]
Loadings > 0.5 kg of BOD/day/kg of MLSS	1.0–1.04	[15]
Trickling filters	1.035	[17]
Aerobic lagoons	1.035	[18]
Aerobic-facultative lagoons	1.07–1.08	[19]
BOD bottle	1.056 (20–30°C)	[32]
	1.135(4–20°C)	[32]

The wide variation in Θ can be rationalized by considering the nature of the process. In the activated sludge process at low loadings, BOD removal and oxidation depend on diffusion of oxygen into the biological flocs, where it is subsequently utilized by the organisms. At low temperatures, a low oxygen-utilization rate permits greater diffusion of oxygen, and consequently a large portion of the floc is aerobic. At high temperatures, the high respiration rate depletes the oxygen rapidly, and only a small portion of the floc is aerobic. It can be assumed that a large mass of organisms at a low respiration rate (winter) achieves the same degree of oxidation as a small mass at a high respiration rate (summer), and hence the coefficient Θ is approximately 1.0.

By contrast, at high organic loadings, the floc tends to become dispersed (bulking sludge), and each organism is more directly affected by changes in temperature. The coefficient, Θ, therefore increases. In aerated lagoons, at a low solids level, the organisms are more dispersed, and the temperature coefficient is high. This is also true for the BOD bottle and the stream. Trickling filters are analogous to activated sludge except that oxygen diffusion is uniplaner into the film. Similar calculations as for activated sludge lead to a coefficient Θ of 1.035. Marked improvement in aerated lagoon operation during the winter months can be achieved by adding a recycle and increasing the solids level in the basin.

NITRIFICATION

The most work on nitrification in recent years has been reported by Downing [20] and by Wuhrmann [21]. Nitrification results from the oxidation of ammonia by *Nitrosomonas* to nitrite and the subsequent oxidation of the nitrite to nitrate by *Nitrobacter*. Since a buildup of nitrite is rarely observed, it can be concluded that the rate of conversion to nitrite controls the rate of overall reaction.

The reactions which occur are as follows. Ammonia is oxidized to nitrite by *Nitrosomonas*:

$$2NH_4^+ + 3O_2 \longrightarrow 2NO_2^- + 2H_2O + 4H^+$$

The nitrite is then oxidized to nitrate by *Nitrobacter*:

$$2NO_2^- + O_2 \longrightarrow 2NO_3^-$$

The nitrifiers are autotropic organisms and use CO_2 or HCO_3 as a carbon source. The organisms can survive with an initial lag period under anaerobic conditions for at least 4 hr.

For effective nitrification to occur, the sludge retention period or sludge age, G, must be greater than the growth rate of the nitrifying organisms. Shorter sludge ages will result in a washout of these organisms. The results of Downing and Wuhrmann show that these resultant retention periods are usually sufficient to effect substantially complete nitrification. The data of Downing have been converted in terms of sludge age to effect nitrification. Nitrification as related to sludge age in sewage oxidation is shown in Fig. 9.9. Temperature will exert a profound effect on nitrification. Downing [22] reported a rate relationship,

$$K_N = 0.18 \cdot 1.128^{(T-15)}$$

A rate K of 0.18 at 15°C was reported for *Nitrosomonas*. The temperature effect on the required sludge age using the data of Downing is shown in Fig. 9.9. Nitrification as related to sludge age is shown in Fig. 9.10.

Percent nitrification as defined in these figures is the percentage of total nitrate formed after complete oxidation. pH

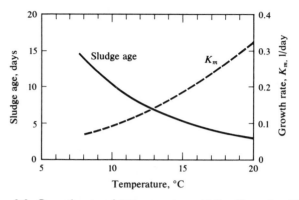

Figure 9.9 Growth rate of *Nitrosomonas*. (After Downing [*20*].)

has a significant effect on the growth rate of both *Nitroso-monas* and *Nitrobacter*. The optimum pH range for *Nitroso-monas* is 7.6 to 8.0, and for *Nitrobacter* 7.8. The rate of nitrification has been shown to be dependent on the dissolved oxygen level at concentrations less than 2.0 mg/l. The rate is independent of ammonia concentration at levels in excess of ap-

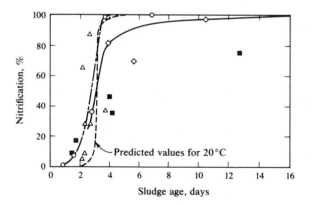

Figure 9.10 Relationship between nitrification and sludge age in the activated sludge process. Data: circles, Wuhrmann [*21*]; diamond, Ford [*13*], laboratory, sewage at 23°C; squares, Balakrishman [*27*], laboratory, acetate and sucrose at 23°C; triangles, conventional sewage-treatment plants, annual operation.

proximately 0.5 mg/l. Heavy metals such as Cr, Ni, and Zn are toxic at concentrations of 0.25 mg/l. Organic carbon may be toxic at a high concentration level.

DENITRIFICATION

The most exhaustive study on denitrification has been reported by Wuhrmann [21]. Many of the heterotropic bacteria present in activated sludge are facultative and can reduce nitrate. Nitrate under these conditions will reduce to N_2 and small quantities of N_2O.

The pH of the mixture profoundly affects the rates of denitrification relative to the presence of dissolved oxygen. pH values in the acid range will permit active denitrification in the presence of dissolved oxygen, whereas strict anaerobic conditions are mandatory at pH levels above 7.0. Since most activated sludge plants operate at or near a neutral pH, anaerobic conditions should be maintained to promote denitrification.

The denitrifying organisms are heterotropic and require an organic carbon source for growth. It is possible, however, to use the endogenous by-products as a food supply for the denitrifiers. Denitrification rates would increase in the presence of available carbon such as supplied by untreated sewage. The rate of denitrification using an endogenous food source is shown in Fig. 9.11. The data of Culp and Slechta [33] obtained from the Tahoe studies are also shown in this figure.

KINETICS OF ANAEROBIC TREATMENT

Anaerobic treatment is employed for the degradation and breakdown of organic solids or for the breakdown of soluble organics to gaseous end products as shown in Fig. 9.12. The volatile acids formed in the fermentation are acetic, propionic, and butyric. For the most of the longer-chain volatile acids, a given species of methane organisms converts the acid to methane, carbon dioxide, and a second volatile acid having a shorter carbon chain. The second volatile acid is then fermented in a similar fashion. Acetic acid is directly converted to CO_2 and CH_4. Thus the overall conversion is the result of two or more reactions.

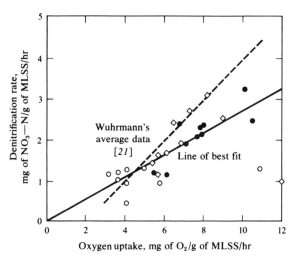

Figure 9.11 Correlation between oxygen uptake and denitrification rate. Data: open circles, Culp and Slechta [33]; diamonds, laboratory data, batch runs; filled circles, laboratory data, continuous run.

Successful anaerobic digestion depends upon maintaining a balance between the various rates of reaction occurring in the digester. Since the rate of methane fermentation must control the overall rate to avoid process failure, further consideration of the rate of this fermentation is important.

To effect methane fermentation, sufficient time must be available in the reactor to permit growth of the organisms or they will be washed out of the system. In a completely mixed flowthrough digestion tank, this means that the detention time in the unit must be greater than the growth rate of the methane organisms. It is significant to note that there are several species of methane organisms active in a digestion system, all having different growth rates. Andrews et al. [4] have shown that some organisms with a high growth rate (< 2 days) can produce methane, probably from the fermentation of formate, methanol, CO_2 and H_2, and possibly some volatile acid fermentation. Other organisms require residence times of up to 20 days. Although data are limited, some results have been reported relative to the growth rate of methane organisms. These data are summarized in Table 9.3.

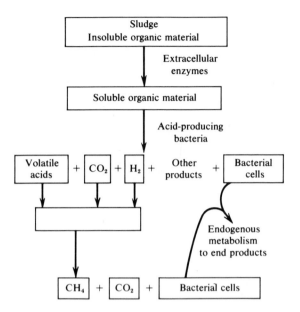

Figure 9.12 Mechanism of anaerobic sludge digestion.

As shown in Fig. 9.13, at low residence times there will be volatile solids reduction as a result of the liquefaction of the solids and the subsequent conversion to volatile acids by acidification. During this period a small amount of methane fermen-

Table 9.3. Growth Rate of Methane Organisms

Substrate	Temp.,°C	Residence Time, days	Reference
Methanol	35	2	[23]
Formate	35	3	[23]
Acetate	35	5	[23]
Propionate	35	7.5	[23]
Primary and			
activated sludge	37	3.2	[24]
Acetate	35	2–4.2	[25]
Acetate	25	4.2	[25]
Propionate	25	2.8	[25]
Butyrate	35	2.7	[25]

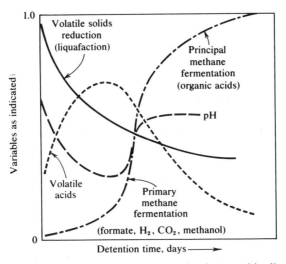

Figure 9.13 Mechanism of continuous mixed anaerobic digestion.

tation may occur (depending on environmental conditions such as pH) primarily due to reduction of formate, methanol, CO_2, and H_2. There will be a decrease in pH and a corresponding increase in the volatile acid concentration. There will be very little COD reduction, however, because the organics have merely been converted from a solid form to a soluble form in the supernatant liquor. When the detention time in the digester exceeds the growth rate of the principal methane organisms there will be a rapid increase in methane production with a corresponding decrease in volatile acid concentration and an increase in pH. There are probably several methane organisms responsible for the volatile acid conversion, each of which will have a different generation time or growth rate; the methane production curve is relatively flat, as shown in Fig. 9.13.

The acetic acid at higher residence times are obtained from two sources: direct fermentation, and the breakdown of higher carbon acids to acetic. The major part of the methane production comes from acetic acid fermentation, although some is generated from the breakdown of the higher acids. At long residence periods, substantially all of the volatile acids are converted to methane and carbon dioxide. It should be noted that since all of the volatile solids present are not degradable in the digestion unit, a portion will remain, even after long periods of retention. For sewage sludge, this fraction is approximately 40 percent.

Gas Production   The major part of the gas produced in a sludge digester comes from the breakdown of volatile acids. Some gas is produced by the early stages of methane fermentation of CO_2 and H_2, methanol, etc., but this contribution is probably very small. The gas will be composed of CH_4, CO_2, and small quantities of H_2S and H_2. The percentage of CH_4 in the gas will depend in large measure on the residence time, the percentage of CO_2 being higher at the lower residence times with corresponding lesser numbers of methane bacteria. Lawrence and McCarty [25] have shown from theoretical considerations supported by experimental evidence that 5.62 ft³ of methane gas will be produced per pound of COD reduced. The reported gas production for volatile solids reduction in a well-operating anaerobic digestion tank is 17 to 20 ft³/lb of VS destroyed with a methane content of about 65 percent. This is about equivalent to 5 to 7 ft³ of CH_4/lb of COD destroyed, which is close to the value reported by Lawrence and McCarty. It is significant to note at this point that these values are a maximum, assuming complete conversion of the solids to methane. Volatile solids reduction can, of course, occur by liquefaction and conversion to volatile acids without any COD reduction. Under these conditions, the methane yield per unit of volatile solids reduction may be very low.

AEROBIC BIOLOGICAL TREATMENT PROCESSES

The various biological treatment processes are summarized below:

1. Activated sludge should provide an effluent with a soluble BOD_5 of less than 10 to 15 mg/l and a total BOD_5, including carryover suspended solids, of less than 20 mg/l. The process requires treatment and disposal of excess sludge and would

generally be considered where high effluent quality is required, available land area is limited, and waste flows exceed 0.1 Mgd.

2. Extended aeration or total oxidation will provide an effluent with a soluble BOD_5 of less than 10 to 15 mg/l and a total BOD_5 of less than 40 mg/l. The suspended solids carryover may run as high as 50 mg/l [high clarity (low solids) effluent will usually require posttreatment by filtration, coagulation, etc.]. This process is usually considered for waste flows less than 2.0 Mgd.

3. Contact stabilization is applicable where a major portion of the BOD is present in colloidal or suspended form. As a general rule, the process should be considered when 85 percent of the BOD_5 is removed after 15 min of contact with aerated activated sludge. The effluent suspended solids are of the same order as those obtained from activated sludge.

4. An aerobic lagoon is only applicable where partial treatment ($\sim$ 50 to 60 percent BOD_5 reduction) and a high effluent suspended solids are permissible. This process should be considered as a stage development, which can be converted into an extended aeration plant at some future date by the addition of a clarifier, return sludge pump, and additional aeration equipment.

5. An aerated lagoon will provide effluent soluble BOD_5 of less than 25 mg/l with a total BOD_5 of less than 50 mg/l, depending on the operating temperature. The effluent suspended solids may exceed 100 mg/l. The system is temperature sensitive and treatment efficiencies will decrease during winter operation. Posttreatment is necessary if a highly clarified effluent is desired. Large land areas are required for the process.

6. High-rate trickling filters will provide 85 percent reduction of BOD for domestic sewage. Roughing filters at high loadings provide 50 to 60 percent BOD reduction from soluble organic industrial wastes.

7. Anaerobic and facultative ponds for industrial waste treatment should only be considered if odors will not cause a nuisance. If high-degree treatment is required, these ponds must be followed up by aerobic treatment (aerated lagoons, activated sludge, etc.).

ACTIVATED-SLUDGE-PROCESS MODIFICATIONS

The activated sludge process can be defined as a system in which the flocculated, biological growths are continuously circulated and contacted with organic wastewater in the presence of oxygen. The oxygen is usually supplied from air bubbles injected into the sludge liquid mixture under turbulent conditions, or from mechanical surface or other types of aeration units. The process involves an aeration step followed by a solids-liquid separation step (sedimentation), from which the separated sludge is recycled back for mixture with the wastewater. In the aeration step, there occur (1) a rapid adsorption and flocculation of suspended colloidal and soluble organics by the activated sludge, (2) progressive oxidation and synthesis of the adsorbed organics and organics continuously removed from the solution, and (3) further aeration resulting in oxidation and dispersion of a sludge particle.

The various stages of the activated sludge process are shown in Fig. 9.14. Let us first consider domestic sewage. The major portion of the BOD is present in suspended and colloidal form and is rapidly removed in a short detention period by flocculation and agglomeration of the suspended BOD and adsorption of the colloidal BOD (time t_1, Fig. 9.14). As shown in Fig. 9.14, these organics then undergo oxidation and synthesis either in the aeration tank (time t_2, Fig. 9.14) or in a separate sludge aeration tank. In the conventional activated sludge process, the suspended, colloidal, and soluble BOD is removed in a detention period of 4 to 6 hr (time t_2). A minimum dissolved oxygen level in the aeration tanks of 0.5 mg/l is necessary to maintain aerobic activity.

If nitrification (biological oxidation of ammonia to nitrate) is to be accomplished, the detention time must be increased to ensure sufficient time for the nitrifying organisms to grow, as shown in Fig. 9.10.

The nitrifying organisms require a higher dissolved oxygen level in the aeration ($>$ 2.0 mg/l) to ensure maximum activity.

After nitrification, if the dissolved oxygen level is permitted to drop to zero, the organisms in the activated sludge mixture will reduce the nitrate to nitrogen gas, as shown in Fig. 9.14.

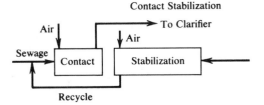

Contact Stabilization

Conventional Activated Sludge

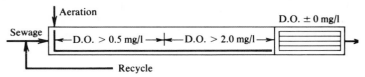

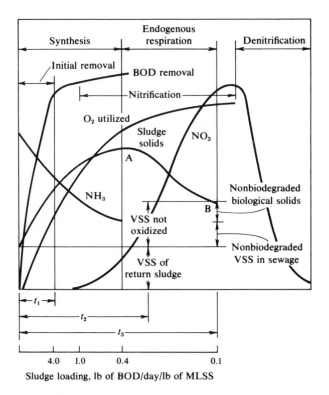

Figure 9.14 Schematic representation of the activated sludge process.

When considering soluble industrial wastes, removal occurs progressively through the aeration tank as shown in Fig. 9.14. This removal is defined by equations (10) and (12a).

The reactions occurring in the activated sludge process can be summarized as follows:

1. Initial removal of suspended and collodial solids by physical agglomeration and flocculation and by adsorption on the biological flocs. These organics are then broken down by biological action in the aeration process.

2. Slow removal of soluble organics from solution by the microorganisms, resulting in oxidation to end products (CO_2 and H_2O) and the synthesis of new microorganisms.

3. Oxidation of ammonia to nitrates by nitrifying organisms. This is a two-stage reaction, ammonia first being oxidized to nitrite, followed by oxidation of the nitrite to nitrate.

4. Oxidation of the biological cells to end products of CO_2, H_2O, NH_3, and phosphorus. A nonbiodegradable residue will remain even after long periods of aeration.

The various modifications of the activated sludge process in use today are as follows.

Conventional Activated Sludge   This process consists of four functional steps: (1) primary sedimentation to remove settleable organic and inorganic solids when they are present in the wastewater, (2) aeration of a mixture of wastewater and a biologically activated sludge, (3) separation of the biologically active sludge from its associated treated liquor by sedimentation, and (4) return of the settled biological sludge to be mixed with raw wastewater (Fig. 9.15).

The conventional activated sludge process can be operated in either of two modes:

1. The plug-flow-type system, in which the wastewater and return sludge are mixed in the head end of the aeration tank. The mixture flows through the tank in a modified plug flow with some longitudinal mixing. This system is shown in Fig. 9.15.

2. Many plants are now being built on the completely mixed principle. In this case, the wastewater and return sludge are returned to the aeration tanks separately and complete mixing occurs in the aeration basin. This system has the advantage of

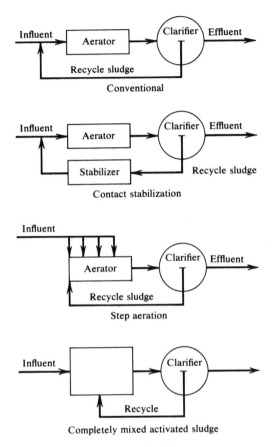

Figure 9.15 Activated sludge processes.

dampening and minimizing slugs and peak concentrations and equalizing variations in the influent wastewater.

In the completely mixed systems, the oxygen-utilization rate is the same in all parts of the basin, allowing aeration equipment to be equally spaced.

Contact Stabilization   When a high percentage of the BOD is removed rapidly by biosorption after contact with well-aerated activated sludge (Fig. 9.15), the contact stabilization process may be advantageously employed. In this process the waste is aerated with stabilized sludge for a short contact period (30 to 60 min). The mixed liquor is then separated by sedimentation and the settled sludge transferred from the clarifier to a sludge stabilization tank, where aeration is continued to complete the oxidization and to prepare the sludge for BOD removal of fresh incoming waste. When the BOD removal rate is too low to attain the desired overall removal in the short contact period, the aeration contact periods must be extended to attain the desired removal. This additional removal is shown in Fig. 9.16. The magnitude of the initial removal is dependent on the characteristics of the sludge and the wastewater. The process is most adaptable to municipal sewage which contains a large percentage of the BOD in suspended form.

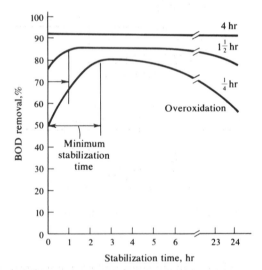

Figure 9.16 Schematic representation of the contact-stabilization process. (From Eckenfelder [36].)

Step Aeration   Step aeration involves the addition of primary sewage effluent along the course of flow of mixed liquor through the aeration tanks to equalize the loading and the oxygen-demand rates through the tank. The return sludge is admitted to the aeration tank influent and the primary effluent added at multiple points along the aerator by means of stop gates (see Fig. 9.15). Aeration detention periods are generally of the order of 2 to 4 hr. The return sludge rates are generally maintained to 25 to 35 percent of the average sewage flow. Operating results compare favorably with conventional treatment. The step aeration process is considered to fall in between the conventional plug-flow-type system and the completely mixed activated-sludge unit.

Modified Aeration Intermediate treatment is attained by modified aeration in which the aeration detention period is about 2 hr based on the average sewage flow plus 10 percent return sludge. Low mixed-liquor solids (200 to 500 mg/l) and a high loading factor (low sludge age) is maintained. BOD removal efficiencies are of the order of 50 to 70 percent domestic sewage treatment. The flow diagram is substantially similar to the conventional process.

Extended Aeration The extended aeration process is based on providing sufficient aeration time for oxidizing the biodegradable portion of the sludge synthesized from the organics removed in the process (time t_3, Fig. 9.14). The extended aeration process is schematically shown in Fig. 9.17. Theoretically, excess sludge in this process is only the nonbiodegradable residues remaining after total oxidation. As sludge is usually wasted from the mixed aeration system, the actual excess is about double the nonbiodegradable residue. The process has been widely applied to the treatment of domestic sewage from small communities, housing developments, and recreational areas and to the treatment of industrial wastes where the daily volume is less than 2.0 mgd. A typical package-type extended aeration plant is shown in Fig. 9.18.

AERATED LAGOONS

An aerated lagoon is a basin of significant depth (6 to 12 ft) (1.8 to 3.6 m) in which oxygenation is accomplished by mechanical or diffused aeration units and from induced surface aeration. Two types can be considered: the aerobic basin, in

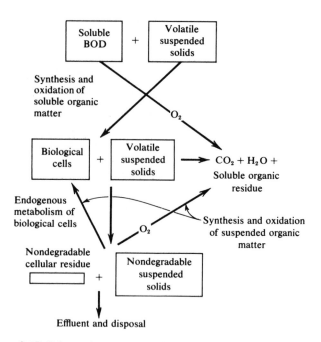

Figure 9.17 Schematic representation of the extended aeration process.

which all solids are maintained in suspension, and the aerobic–anaerobic basin, in which the turbulence level maintained in the basin ensures distribution of oxygen throughout the basin but is usually insufficient to maintain solids in suspension.

In the aerobic basin all solids are in suspension so that the effluent suspended solids will be equal to the solids in the basin. In most cases, separate sludge settling and disposal facilities are required. The aerobic basin is readily adaptable to modification to the activated sludge process.

In the aerobic–anaerobic basin, the major portion of inert suspended solids and nonoxidized biological solids settle to the bottom of the basin, where they undergo anaerobic decomposi-

---

*In aerated lagoons where most of the solids settle, equation (12a) is frequently expressed $s_a - s_e/t = Ks_e$.

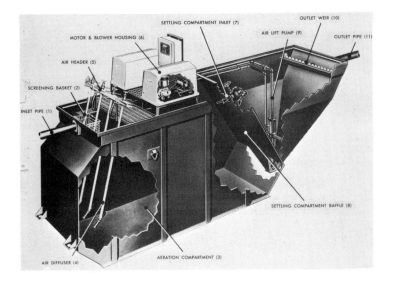

Figure 9.18 Package extended aeration plant. (Courtesy of Smith and
          Loveless, Inc.)

tion. The basin can be modified to include a separate sedimentation compartment to yield a more highly clarified effluent.

These basin types are shown in Fig. 9.19. An aerated lagoon
is shown in Fig. 9.20.

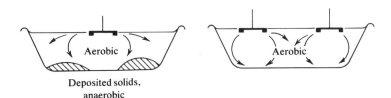

Figure 9.19 Types of aerated lagoons.

BOD-Removal Characteristics Just as in the completely mixed activated sludge process, BOD removal is primarily a function of detention time, the biological solids con-

Figure 9.20 Aerated-lagoon treatment of chemical wastes. (Courtesy of Wells Products.)

centration, the temperature, and the nature of the waste. The removal can be formulated according to equation (12a).

The effluent BOD from the lagoon will be the sum of the soluble BOD remaining as defined plus the BOD contributed by the respiration of suspended solids present in the effluent. (In the aerobic basin this is equal to the solids in the basin, while in the aerobic–anaerobic basin it is the carryover solids not settled in the pond.) In general, for a high-degree-treatment pond, the BOD contribution from the suspended solids for a soluble waste will be approximately 0.3 mg of O_2/mg of SS. The effluent BOD will be

$$s(\text{effluent}) = s + 0.3 \text{ (mg/l of SS)}$$

The removal of BOD from several wastes is summarized in Table 9.1.

Since the solids level maintained in the lagoon is low, temperature variation will exert a profound effect on the rate of BOD removal, as shown on page 169.

Solids Levels in Lagoons   In the aerobic lagoon the solids level will be determined by the solids in the influent waste, the synthesis from the BOD removed, and autooxidation of biological solids. A material balance around the basin shows

$$\text{solids in} - \text{solids out} = \text{solids buildup}$$

and

$$\frac{S_0 + as_r}{1 + bt} = X_v \tag{16}$$

For soluble wastes, X_v will be approximately one half the concentration of BOD applied to the system.

In the aerobic–anaerobic lagoon, the solids level will depend on the mixing regime, which in turn dictates what portion of the solids will settle in the basin.

Oxygen Requirements   In the aerobic lagoon, oxygen requirements can be computed in the same manner as for activated sludge [equation (3)]. In the aerobic–anaerobic lagoon the quantity of biological solids is maintained at a low level and BOD is fed back to the liquid from the anaerobic fermentation of the settled solids, so oxygen requirements can be related to the BOD removal:

$$\text{lb of } O_2/\text{day} = a'' \cdot \text{lb of BOD removed/day} \tag{17}$$

Results obtained on various industrial wastes indicate that the coefficient a'' will vary from 0.9 to 1.4. For paper mill wastes varying in BOD from 150 to 400 mg/l, the oxygen-uptake rate has been observed to vary from 1.0 to 4.0 mg/l/hr.

Aeration   Oxygen is transferred to an aerated lagoon from mechanical or diffused aeration units and from surface aeration resulting from velocity gradients induced by the aeration device. In most cases 10 to 20 percent of the total oxygen required results from surface aeration.

Mechanical surface aeration units can be installed on a float or raft arrangement or on a permanent base. The spacing of these units should permit sufficient mixing to ensure uniform dissolved oxygen throughout the basin. Present information indicates that spacings up to 400 ft (122 m) at power levels of 0.015 to 0.02 hp/1000 gal (0.0030 to 0.0041 kw/m³) of basin volume will accomplish this. These power levels will usually not maintain solids in suspension.

Temperature Effects The liquid temperature in the aerated lagoon will depend upon the heat balance resulting from changes in ambient temperature and waste temperature. Heat is lost through evaporation, convection, and radiation and is gained by solar radiation. It has been found that heat loss by evaporation and heat gain by solar radiation are small compared to the losses by convection and radiation and can usually be neglected.

The heat loss by convection can be computed from the relationship

$$H_e = (0.8 + 0.32W/2)(T_w - T_a) \tag{18}$$

in which T_w and T_a are the lagoon and air temperature, respectively, and W is the mean wind velocity.

The heat loss by radiation can be expressed by the relationship

$$H_r = T_w - T_a \tag{18a}$$

Combining equation (10) and (11) and using a mean wind velocity and surface-turbulence correction, the heat balance for the aerated lagoon can be expressed

$$fA(T_w - T_a) = Q(T_i - T_o) \tag{18b}$$

in which A is the lagoon surface area, Q the waste flow, and T_i and T_o the influent and effluent temperature to the lagoon, respectively. In a completely mixed basin, $T_o = T_w$ can be reexpressed to define the mean basin temperature:

$$T_1 - T_w = \frac{(T_w - T_a)fA}{Q} \tag{18c}$$

The coefficient f is a proportionality factor containing the heat-transfer coefficients, the surface-area increase from the aeration equipment, and wind and humidity effects.

In equation (18c), when Q is in mgd and A in square feet, f has an approximate value of 12×10^{-6} for the central portion of the United States [19]. In computing the estimated lagoon temperature, local weather bureau records are employed to define T_a.

AEROBIC BIOLOGICAL TREATMENT DESIGN

Effective biological treatment requires pretreatment for removal of contaminants which would cause operating problems or upset in the biological process. These can be summarized:

Pretreatment or Primary Treatment
1. If suspended solids are present in excess of approximately 125 mg/l, solids separation by lagooning, sedimentation, or flotation should be considered. For estimating purposes only, a sedimentation tank would have an overflow rate of about 1000 gpd/ft² (4080 m³/day/m²).

2. If oil, grease, or flotables exceed 50 mg/l, a skimming tank or separator should be provided.

3. Heavy metals (Cu, Zn, Ni, and so on) should be removed prior to biological treatment.

4. a. If the pH exceeds pH 9.0, neutralization should be provided if the ratio of caustic alkalinity (expressed as $CaCO_3$) to COD removed exceeds 0.7 lb of $CaCO_3$/lb of COD or 0.56 lb of $CaCO_3$/lb of BOD_5 removed. Neutralization need only reduce the alkalinity to the aforementioned levels.

   b. In some cases where a wide variation in alkalinity is encountered during the day or plant operating schedule, the aforementioned levels can be achieved by providing an equalization tank or pond.

5. If the waste contains organic acids, biooxidation will convert these acids to CO_2 and bicarbonate salts, provided the process design reduces these to < 25 mg/l as BOD_5.

6. When mineral acids are present, neutralization or equalization should be provided if the pH is less than 4.5.

7. Sulfides should be prestripped or otherwise removed if their concentration exceeds 50 mg/l.

8. If the influent BOD loading in lb/day, based on 4-hr composites, exceeds a 3:1 ratio, an equalization tank should be considered to bring the variation within this range.

After selection of the possible consideration, the preliminary design calculations should be developed as follows. This procedure presumes that no pilot-plant or laboratory-scale treatment study data are available. If such data are available, the appropriate coefficients and factors developed from the laboratory study should be used.

General

1. Select the appropriate k, a, a', b, and b'. It should be recognized that at this time there is limited data on the treatment of many industrial wastes, so some of the coefficients are necessarily estimations. As more data are accumulated, these coefficients will be refined.

2. Each of the processes are designed using the equations shown in Table 9.4. It is important to remember that the coefficients generally apply to all systems, and the principle differences between systems are the changes in concentration of biological solids and retention time with the exception of the aerated lagoon, which has a feedback of BOD from the anaerobic sublayer.

Aerobic Lagoons

1. The required retention period for a specified effluent soluble BOD is estimated by combining equations (12a) and (16).

$$t, \text{days} = \frac{1}{24\, aks_e - b}$$

2. The equilibrium volatile solids in the lagoon is computed from equation (16):

$$X_v = \frac{S_0 + as_r}{1 + bt}$$

and the total suspended solids is

$$X_a = \frac{X_v}{\text{volatile fraction}}$$

Table 9.4. Summary of Design relationships

The basic relationships applicable to the design of biological waste treatment facilities are summarized below.

1. BOD or COD removal in a completely mixed basin, aerobic or anaerobic:

$$\frac{s_a - s_e}{X_v t} = k s_e \tag{12a}$$

2. Sludge yield:

$$\Delta X_v = S_0 + a s_r - b X_v \tag{1}$$

(conventional systems; aerobic or anaerobic).

3. Equilibrium solids in an aerobic lagoon or anaerobic flowthrough system:

$$\frac{S_0 + a s_r}{1 + bt} = X_v \tag{16}$$

4. Sludge age:

$$G = \frac{X_v}{\Delta X_v} = \frac{X_v}{a s_r - b X_v} \tag{14}$$

5. Oxygen requirements (aerobic systems):

$$O_2 = a' s_r + b' X_v \text{ (conventional systems)} \tag{3}$$
$$O_2 = a' s_r + b' X_d \text{ (total oxidation systems)} \tag{3a}$$

6. Nutrient requirements:

$$\text{Nitrogen (N)} = 0.12 \, \Delta X_v + 1.0 \text{ mg/l} \tag{19}$$
$$\text{Phosphorus (P)} = 0.02 \, \Delta X_v + 0.5 \text{ mg/l} \tag{19a}$$

7. Organic loading:

$$\frac{F}{M} = \frac{24 \, s_a}{X_v t} \tag{20}$$

For soluble organic waste the volatile fraction can be assumed to be 80 percent. When suspended solids are present in the waste, this estimate should be revised, depending on the volatile fraction of the influent suspended solids.

3. The effluent BOD will be composed of the soluble BOD plus that contributed by the concentration of suspended solids in the basin, which will be approximately the same as in the effluent for a completely mixed system:

$$\text{effluent BOD} = s_e + fX_a$$

when f is the BOD contributed per unit of effluent suspended solids. This is selected from Fig. 9.21 or from a plot developed from experimental data.

4. The oxygen requirements are calculated from equation (3) and include the oxygen used by the organisms for growth and those used for endogenous respiration.

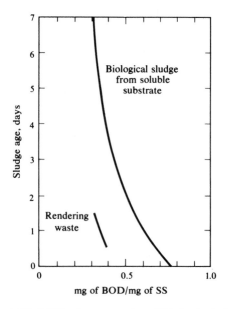

Figure 9.21 BOD characteristics of biological sludges.

Aerated Lagoons

1. Since the relatively low power level in the system permits the major portion of the suspended solids to settle to the bottom of the basin, the solids level in suspension will be low. The volatile solids maintained in suspension will vary from 50 to 150 mg/l depending on the power level maintained. The effluent soluble BOD is selected and the required retention period calculated from equation (12a):

$$t = \frac{s_r}{24kX_v s_e}$$

2. The total effluent BOD is computed as in step 2 under Aerobic Lagoons using the effluent volatile suspended solids selected in step 1.

3. Owing to the feedback of BOD from the anaerobic sublayer, the oxygen requirements will be more than those calculated from equation (3). The oxygen requirement will be 1.2 to 1.4 times the BOD_5 removed from the waste.

4. It is important that enough power is supplied to maintain uniform distribution of oxygen throughout the basin, particularly when the waste is low in BOD.

Activated Sludge Process

In the activated sludge process, a primary requirement for effective operation is rapid settling and compaction of the sludge. This means that the sludge loading, lb of BOD/day/lb of MLSS must fall within the flocculation ranges and below the level conducive to the growth of filamentous organisms (see Fig 9.9). The activated sludge process is schematically shown in Fig. 9.14, which includes all ranges of operation. The design relationships for the various activated sludge modifications are summarized as follows:

1. Conventional activated sludge
   a. Required basin volume:

$$F/M = \frac{24s_a}{X_v t} \tag{20}$$

F/M must fall in flocculation range (usually 0.3 to 0.7 lb of BOD/day/lb of MLSS); X_v is selected based on sludge settling properties; usually between 2000 and 4000 mg/l will yield optimum removal.

$$V = \frac{Q}{24t}$$

where Q is the waste flow in V, the basin volume in M gal.
　　b. Oxygen requirements—equation (3)
　　c. Sludge yield—equation (1)
　　d. Nutrient requirements—equation (19)
　2. Conventional activated sludge with nitrification
　　a. Required basin volume—equation (14), with sludge age selected from Fig. 9.9.
　　b. Oxygen requirements—equation (3) plus N oxidized
　　c. Sludge yield—equation (1)
　3. Extended aeration
　　a. Required basin volume

$$X_a = \frac{a_0 s_r}{fb \ (\text{vol. fract.})}$$

where a_0 = fraction of BOD converted to degradable solids
　　f = fraction of degradable solids in system

$$V = \frac{X_a \ (\text{lb})}{X_a \ (\text{mg/l}) \cdot 8.34}$$

in which V = M gal of basin volume, or

$$V = \frac{X_a \ (\text{kg}) \cdot 10^2}{X_a \ (\text{mg/l})}$$

in which V = m³ of basin volume.
　　b. Oxygen requirements—equation (3a).
　4. Design—activated sludge
　　a. It is first necessary to assume a mixed-liquor suspended-solids concentration; solids in the return

sludge and the return sludge rate. In general, good settling flocculated sludge can be expected to concentrate to 10 to 12,000 mg/l; in the final settling tank with
return sludge rates of 20 to 50 percent of the average
waste flow. Suggested maximum return sludge rate is
150 percent of the average flow rate. This would indicate an MLSS selection of 3000 mg/l as a conservative operating level.

b. Since a well-flocculated good settling sludge is essential to effective process operation, the loading level
(lb of BOD/day/lb of MLSS) must be in the proper
range (see Fig. 9.7). For most wastes, a value 0.5
based on average waste loadings will yield a conservative design. The required detention time should be calculated from both the loading level and from equation
(12a). The largest value should be used for design. For
a highly degradable waste such as sugar, or brewery,
the maximum loading, flocculation will probably control, while for a complex waste such as petrochemical,
the effluent quality will most likely control the required detention period.

Extended Aeration

1. In the extended aeration process, the retention time or
basin volume required is that necessary to oxidize all of the
degradable sludge produced in the system by synthesis. For
a soluble waste, the quantity of sludge that must be maintained
under aeration to accomplish this oxidation is

$$X_{avg} = \frac{a_0 S_r}{fb}$$

in which a_0 is the fraction of BOD converted to degradable
sludge and can be assumed to be equal to $(0.88a)$ and S_r the
lb/day or kg/day of BOD removed. The total solids under aeration is then

$$\frac{X_{avg}}{\text{fraction volatile}} \qquad X_{avg} = \text{average mixed-liquor VSS, mg/l}$$

The volatile fraction for soluble wastes in an extended aeration process will probably average 70 percent. Assuming 3500 mg/l mixed-liquor suspended solids, the required aeration volume is

$$V = \frac{X_a \text{ (lb)}}{3500 \times 8.34} \quad \text{(M gal)} \quad V = \frac{X_a \text{ (kg)} \cdot 10^2}{3500}$$

2. The aeration requirements are calculated from equation (3a).

3. The excess sludge will be the nonbiodegradable residue and will be approximately equal to

$$\Delta X_v = 0.23 a s_r - \text{effluent loss}$$

In a continuous completely mixed extended aeration plant with intermittent sludge wasting from the final clarifier, both degradable and nonbiodegradable sludge will be wasted from the mixture. This will result in a total wastage about twice that calculated from Step 3.

TRICKLING FILTERS

A trickling filter is a packed bed of media covered with slime over which wastewater is passed. Oxygen and organic matter diffuse into the film where oxidation and synthesis occur. End products (CO_2, NO_3, etc.) counterdiffuse back into the flowing liquid and appear in the filter effluent as shown in Fig. 9.22. A filter is shown in Fig. 9.23.

In a manner analogous to activated sludge, under plug-flow conditions, BOD removal is related to the available biological slime surface and to the time of contact of wastewater with that surface. The general relationship developed for the activated sludge process is

$$\frac{s}{s_0} = e^{-kX_v t} \tag{10a}$$

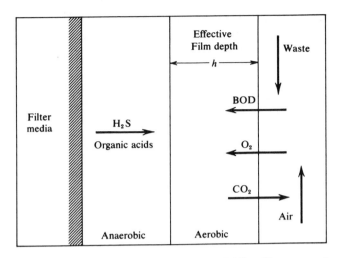

Figure 9.22 Schematic representation of trickling filter operation.

In a trickling filter the mean time of contact has been found to be

$$t = \frac{CD}{Q^n}$$

(21)

in which C and n are characteristic of the filter media, $D =$ depth, and $Q = $ flow.

Figure 9.23 Trickling filter plant for municipal sewage. (Courtesy of Eimco, Inc.)

The biological solids concentration in a filter is proportional to the specific surface, A_v, or

$$X_v \sim A_v$$

Equation (10) can then be modified for a trickling filter:

$$\frac{s}{s_0} = e^{-kA_v{}^m D/Q^n} \qquad (22)$$

For a specific packing where A_v is constant, equation (20) can be expressed

$$\frac{s}{s_0} = e^{-KD/Q^n} \qquad (23)$$

Filter packings have been composed of $2\frac{1}{4}$ to 4 in. (5.5 to 10 cm) of rock (9 to 25 ft²/ft³) with depths of 3 to 8 ft (1 to 2.5 m) and more recently plastic packing (20 to 35 ft²/ft³) with depths up to 40 ft (12 m). As BOD removal results in slime growth, a maximum A_v of about 30 ft²/ft³ is recommended for the treatment of carbonaceous wastes to avoid filter plugging and ponding.

In many cases recirculation of filter effluent will improve the overall BOD removal through the filter. [This is particularly true when the influent BOD is high, > 500 mg/l, and dilution is required to maintain aerobic conditions in the filter and where n in equation (23) is less than 0.67.] When recirculation is employed, equation (23) is modified to consider dilution of the filter influent:

$$\frac{s}{s_a} = \frac{e^{-KD/Q^n}}{(1 + N) - Ne^{-KD/Q^n}} \qquad (24)$$

in which N is the recirculation ratio R/Q.

Filter performance is influenced by temperature. In general, the BOD removal rate, K, is affected by temperature in accordance with the relationship

$$K_t = K_{20°C} \cdot 1.035^{(T - 20)} \qquad (25)$$

Applications of Trickling Filters to Wastewater
Treatment   In the treatment of domestic sewage, filter performance is influenced both by the specific surface and by the initial BOD of the sewage. (Increasing the initial BOD will usually result in an increase in suspended and colloidal organics which are flocculated through the filter, yielding a high rate of removal.) A relationship has been found from the correlation of data from various sources [26] corrected to 20°C:

$$\frac{S}{S_0} = e^{-0.00362 A_v^{0.644} S_0^{0.540} D/Q^n}$$

These data are shown in Fig. 9.24.

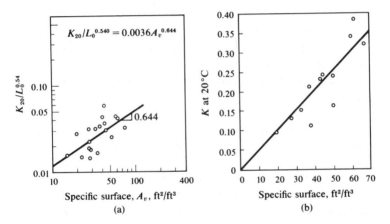

Figure 9.24   Relationships between specific surface and removal rate coefficient: (a) domestic-sewage BOD and (b) nitrification.

The removal-rate coefficient K for soluble industrial waste is low, and hence filters are not economically attractive for high efficiency (> 85 percent BOD reduction) of such wastes. Plastic packed filters have been employed, however, as a pretreatment method for high-BOD wastes. BOD removals of 50 to 60 percent have been achieved at hydraulic and organic loadings of > 4 gpm/ft² (160 l/min/m²) and > 500 lb of BOD/day/1000 ft³ (kg of BOD/day/m³) at filter depths of up to 40 ft

(12 m). The filter effluent requires additional aerobic treatment for high removal levels.

Nitrification can be achieved through a trickling filter at low loadings. Since slime growth from nitrification is very low (see p. 155), high specific surfaces can be employed if the organic carbon content of the influent waste is low. Nitrification through a trickling filter is shown in Fig. 9.25 [27].

PROCEDURE FOR THE DESIGN OF A TRICKLING FILTER

Data to Be Collected
1. Volume and characteristics of waste.
2. Determine whether recirculation is necessary.

Pretreatment or Primary Treatment
1. The BOD applied to the filter should generally not exceed 500 mg/l to ensure aerobic conditions. Recirculation should be employed when wastes have BOD values in excess of this value.

2. When a filter packing with a high specific surface is used, a minimum hydraulic loading is essential to avoid the accumulation of excessive biological growths and resulting filter ponding.

3. The pH of the wastewater applied to the filter should be such that the excess alkalinity or acidity is neutralized in the top few feet of the filter media. Excessive acidity or alkalinity can frequently be neutralized by recirculation of filter effluent.

Design Procedure A
1. Collect data to develop a relationship among BOD removal, depth, and hydraulic loading. At least three different hydraulic loading rates should be tried.

2. Develop a plot of depth vs. percent BOD remaining for the different hydraulic loading rates (Fig. 9.26a).

3. Determine the slopes of the lines in Fig. 26a.

4. Plot Q vs. slopes (on log-log paper) and determine n (Fig. 26b).

5. Develop a plot of D/Q^n vs. percent BOD remaining; the slope of the line gives the coefficient K (Fig. 26c).

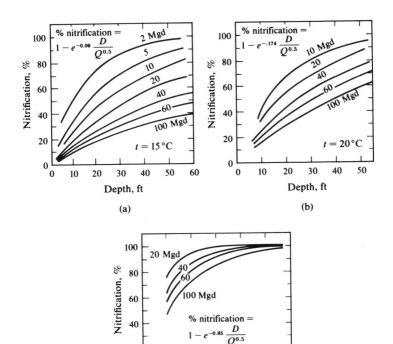

Figure 9.25 Predicted performance characteristics curves for nitrification in trickling filters at different temperatures. $A_v = 70$ ff²/ff³ and $n = 0.5$. (From Balakrishnan [27].)

6. The design equation for the %   BOD removal is given by

$$\% \text{ BOD removal} = 1 - e^{-KD/Qn}$$

7. Selecting a suitable depth, determine the flow rate necessary for a given percent BOD removal.

8. From the flow rate and the total volume of flow to be treated, determine the size of the filter.

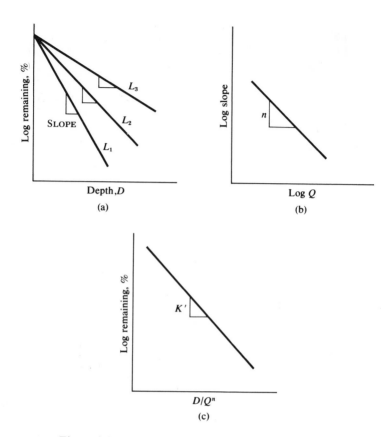

Figure 9.26 Trickling filtration design relationships.

Design Procedure B

1. In cases where experimental data are not available, a first estimate of filter requirements can be made using the coefficients listed in Table 9.5 with equations (23) and (24).

2. Select a filter depth; for rock filters, depths will usually vary from 3 to 10 ft (1 to 3 m), while for plastic packing, depths of up to 40 ft (12 m) have been employed.

3. Using equation (23) compute the hydraulic loading to the filter; an alternative design might consider recirculation as calculated from equation (24).

Table 9.5. Design-Procedure Coefficients

Waste	Filter Media	ft²/ft³	m²/m³	K	n
Sewage	Surfpac	28.0	91.9	0.079	0.5
Fruit canning	Surfpac	28.0	91.9	0.0177	0.5
Boxboard	Surfpac	28.0	91.9	0.0197	0.5
Steel coke plant	Surfpac	28.0	91.9	0.0211	0.5
Textile	Surfpac	28.0	91.9	0.0156	0.5
Textile	Surfpac	28.0	91.9	0.0394	0.5
Textile	Surfpac	28.0	91.9	0.0268	0.5
Pharmaceutical	Surfpac	28.0	91.9	0.0292	0.5
Slaughterhouse	Surfpac	28.0	91.9	0.0246	0.5

4. The total filter area required is computed:

$$\text{area (acres)} = \frac{\text{wastewater flow including recirculation (Mgd)(m}^3/\text{day)}}{\text{hydraulic loading (Mgd)(m}^3/\text{aq. m/day)}}$$

5. If the filter is used as a roughing filter with high organic loadings (50 to 60 percent BOD removal), the same procedure (steps 1 to 4) is followed.

ANAEROBIC DIGESTION

The anaerobic conversion of organic solids to inoffensive end products is very complex, and is the result of many reactions, as shown in Fig. 9.12. In the conventional high-rate digestion system, all the reactions shown in Fig. 9.12 occur simultaneously in the same tank. Under equilibrium operating conditions (steady state) all the reactions must be occurring at the same rate, because there is no buildup of intermediate products. Although many factors, such as sludge composition and concentration, pH, and temperature mixing, influence the reaction rates, it is generally assumed that the overall rate is controlled by the rate of conversion of volatile acids to methane and carbon dioxide. Digester upset and failure occurs when there is an imbalance in the rate mechanism, resulting in a buildup of intermediate volatile acids. The optimum conditions to maintain methane fermentation are summarized in Table 9.6.

Table 9.6. Environmental Conditions for Methane Fermentation

Variable	Optimum	Extreme
pH	6.8–7.4	6.4–7.8
Oxidation–reduction potential, mv	-520 to -530	-490 to -550
Volatile acids, mg/l, as acetic	50–500	> 2000
Total alkalinity, mg/l, as CaCO$_3$	1500–5000	1000–3000
Salts		
NH$_4$, mg/l, as N		3000
Na, mg/l		3500–5500
K, mg/l		2500–4500
Ca, mg/l		2500–4500
Mg, mg/l		1000–1500
Gas production, ft^3/lb of VS destroyed	17–22(1060–1370 l/kg of VS)	—
Gas composition, % CH$_4$	65–70	—
Temperature, ^{0}F	90–100	—

The available data would indicate that effective digestion should be possible with detention periods as low as 5 days provided that other environmental conditions are maintained as shown in Table 9.6. Under these conditions, about 70 percent of the degradable organic solids will be liquefied, with the major portion being converted to gas. Increasing the detention time to 10 days should result in 90 percent of degradable organics being liquefied, with over 90 percent of the degradation by-products being converted to gas. These data are graphically shown in Fig. 9.27. Although methane fermentation should be related only to sludge age or digester detention time (as related to methane organism growth rate), Sawyer [31] has shown that at solids feed levels in excess of 8 percent, methane fermentation decreases. This is probably due to the mixing problems encountered with the high solids levels. Grease reduction, however, improves with a higher solids concentration in the feed sludge. A digestion system is shown in Fig. 9.28.

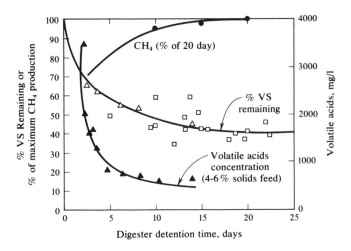

Figure 9.27 High-rate anaerobic digester performance. Data: triangles, Torpey [24]; squares, Estrada [30]; Sawyer [31].

PROCEDURE FOR THE DESIGN OF ANAEROBIC PROCESSES

Data to Be Collected

1. Waste flow, (Mgd) (m³/day), and fluctuations
2. Waste characteristics
   a. Suspended solids and volatile suspended solids, mg/l
   b. COD or TOC, mg/l
   c. Nitrogen (total Kjeldahl nitrogen and organic nitrogen), mg/l
   d. Phosphorus, mg/l
   e. Alkalinity or acidity, mg/l
   f. Presence of heavy metals, oils and toxics, etc.

Pretreatment or Primary Treatment

1. If suspended solids are present in excess of approximately 250 mg/l, solids-separation devices such as a sedimentation or flotation may be provided.

2. If oil, grease, or flotables exceed 100 mg/l, a skimming tank or separator should be provided.

Figure 9.28 Two sludge digestion systems including external heater and gas recirculation for digester mixing. (Courtesy of Pacific Flush Tank Co.)

3. Heavy metals (Cu, Zn, Ni, Pb) should be removed prior to biological treatment.

4. Neutralization of excess alkalinity or acidity should be provided; pH control is critical in the anaerobic process.

Design of the Anaerobic Process

1. Estimate the coefficients a, b, and k of the process for the waste from pilot-plant or laboratory-scale studies (Fig. 9.29).

2. Combining the sludge-growth equation and substrate-removal equation, get a relationship among t, k, a, b, and s_e as follows:

Sludge-growth equation:

$$\frac{s_0 - s_e}{X_a} = \frac{1}{a} + \frac{b}{a}t$$

Substrate-removal equation:

$$\frac{s_0 - s_e}{X_a t} = ks_e$$

or

$$\frac{s_0 - s_e}{X_a} = ks_e t$$

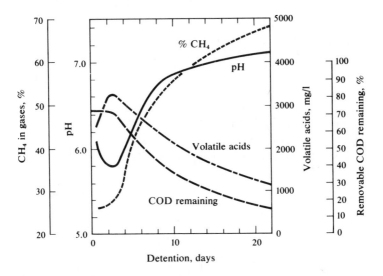

Figure 9.29 Anaerobic degradation of organic waste. (After Andrews et al. [4].)

3. Determine the detention period from the above equation.
4. Calculate X_a of the system using the relationship

$$X_a = \frac{a(s_0 - s_e)}{1 + bt}$$

LAGOONS AND STABILIZATION BASINS

Lagoons and stabilization basins are the most common method of treatment where large land areas are available and a high-quality effluent is not required at all times. There are several types of basins used, depending on the nature of biological activity involved.

Aerobic Algae Ponds Aerobic algae ponds depend upon algae to provide sufficient oxygen to satisfy the oxygen demand exerted by the applied BOD. Because sunlight is essential to algae production, the depth of the pond is limited to

6 to 18 in. (15 to 45 cm). The amount of oxygen produced can be estimated:

$$O_2 = 0.25 FS \tag{26}$$

where O_2 = oxygen production, lb/acre/day (kg/ha/day)
F = light-conversion efficiency, %
S = light intensity, cal/cm^2/day
If the light-conversion efficiency is estimated as 4 percent, then

$$O_2 = S \tag{26a}$$

S may be expected to vary from 100 to 300 cal/cm^2/day during summer and winter for latitude 30°. This in turn would imply maximum loadings of 100 to 300 lb of BOD_u/acre/day (112 to 336 kg BOD_5/day). Mixing the contents a few hours each day is essential to maintain aerobic conditions throughout the basin.

Facultative Ponds The facultative pond is divided into an aerobic layer at the top and an anaerobic layer on the bottom. The aerobic layer is generated by algae which produce oxygen by photosynthesis. The typical green algae in waste stabilization ponds are *Chlamydomonas, Chlorella,* and *Euglena.* Common blue-green algae are *Oscillatoria, Phormibium, Analystis,* and *Anabaena.* The aerobic layer will exhibit a diurnal variation in oxygen content decreasing during the night. Bottom-sludge deposits will undergo anaerobic decomposition, producing methane and other gases. These reactions are shown in Fig. 9.30. Pond depths will vary from 3 to 8 ft (1 to 2.5 m). The effluent from a pond may be expected to increase in the spring as a result of the winter accumulation of deposited sludge and a pond turnover, as shown in Fig. 9.31. In municipal sewage treatment, treatment performance can be predicted by

$$\frac{s}{s_0} = \frac{1}{1 + Kt} \tag{27}$$

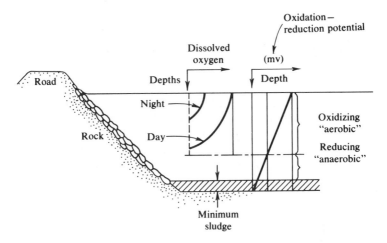

Figure 9.30 Waste-stabilization pond, faculative type. (After Gloyna [28].)

(The coefficient K has a temperature coefficient, Θ, of 1.085.)

Anaerobic Ponds Anaerobic ponds are loaded to such an extent that anaerobic conditions exist throughout the liquid volume. The biological process is the same as that occurring in anaerobic digestion, being organic acid formation followed

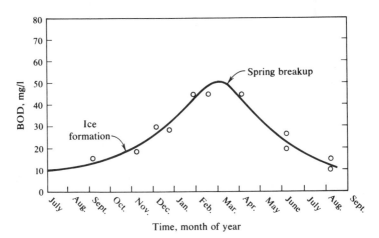

Figure 9.31 Typical seasonal variation of waste-stabilization-pond effluent BOD (Canada). (After Fisher et al. [29].)

by methane fermentation. High organic loading (500 to 2000 lb of BOD/acre/day) (560 to 2244 kg of BOD/ha/day) is usually employed, resulting in removal efficiencies of the order of 70 percent. Since this degree of efficiency is usually not adequate for discharge into a receiving water, anaerobic ponds are usually followed by aerobic ponds. Gloyna [28] has determined the area ratio between anaerobic and aerobic ponds to be between 10:1 and 5:1. Ponds with a ratio less than 3:1 will be sensitive to short-term changes in BOD. Deep depths in anaerobic ponds are desirable to provide maximum heat retention.

Certain design considerations are important to the successful operation of stabilization basins (see Table 9.7). These have

Table 9.7.   **Design Criteria for Stabilization Basins**

	Aerobic[a]	Facultative	Anaerobic	Aerated
Depth, ft	0.6–1.0	3–8	8–15	8–15
m	0.2–0.3	1–2.5	2.5–5	2.5–5
Detention, days	2–6	7–50	5–50	2–10
BOD Loading				
lb/acre/day	100–200	20–50	250–4000	—
kg/ha/day	111–222	22–55	280–4500	
Percent BOD				
removal	80–95	70–95	50–80	80–95
Algae concn.,				
mg/l	100	10–50	Nil	Nil

[a] Must be periodically mixed; velocity 1 to 1.5 fps.

been discussed by Gloyna [28]. Embankments should be constructed of impervious material with maximum and minimum slopes of 3 to 4:1 and 6:1, respectively. A minimum freeboard of 3 ft should be maintained in the basin. Provision should be made for bank protection from erosion. Wind action is important for pond mixing and will be effective with a fetch of 650 ft (200 m) in a pond with a depth of 3 ft (1 m).

References

1. P. L. McCarty, "Thermodynamics of Biological Synthe-

sis and Growth," *Advances in Water Pollution Research,* Vol 2, Pergamon Press, 1964.

2. J. A. Servizi and R. H. Bogan, "Free Energy as a Parameter in Biological Treatment," *Proc. ASCE,* **89,** No. SA3, 17 (1963).

3. W. W. Eckenfelder and D. J. O'Connor, *Biological Waste Treatment,* Pergamon Press, Oxford, 1961.

4. J. F. Andrews et al., *Kinetics and Characteristics of Multistage Methane Fermentations,* SERL Rep. 64-11, University of California, Berkeley, 1962.

5. K. Wuhrmann, *Hauptwirkungen und Wechsel Wirk Kungen einiger Betriebsparameter im Belebschlammsystem Ergebnissemehrjahirger,* Grossversuche Verlag, Zurich, 1964.

6. B. J. McCabe and W. W. Eckenfelder, "BOD Removal and Sludge Growth in the Activated Sludge Process," *J. Water Pollution Control Federation,* **33,** 258–271 (1961).

7. I. S. Wilson, "Concentration Effects in the Biological Oxidation of Trade Wastes." *Proc. 1st Intern. Conf. Water Pollution Research,* London, 1962, Pergamon Press.

8. K. Wuhrmann, in *Biological Treatment of Sewage and Industrial Wastes,* Vol. I, J. McCabe and W. W. Eckenfelder, eds., Reinhold, New York, 1956.

9. L. Tischler and W. W. Eckenfelder, "Linear Substrate Removal in the Activated Sludge Process," *Proc. 4th Intern. Conf. Water Pollution Research,* Prague, 1968, Pergamon Press.

10. A. F. Gaudy et al., "Sequential Substrate Removal in Heterogeneous Population," *J. Water Pollution Control Federation,* **35,** No. 7, 903 (1963).

11. A. Pasveer, "Distribution of Oxygen in Activated Sludge Floc," *Sewage Ind. Wastes,* **26,** No. 1, 28 (1954).

12. D. Okun, in *Advances in Biological Waste Treatment,* Pergamon Press, Oxford, 1963, p. 38.

13. D. L. Ford, Ph.D. Thesis, University of Texas, Austin, Texas, 1966.

14. R. P. Logan and W. E. Budd, in *Biological Treatment of Sewage and Industrial Wastes,* Vol. I, J. McCabe and W. W. Eckenfelder, eds., Reinhold, New York, 1956.

15. K. Wuhrmann, "High Rate Activated Sludge Treatment and Its Relation to Stream Sanitation," *J. Water Pollution Control Federation*, **26**, 1 (1954).

16. W. Von Der Emde, in *Advances in Biological Waste Treatment,* Pergamon Press, Oxford, 1963.

17. W. Howland, "Flow Over Porous Media as in a Trickling Filter," *Proc. 12th Ind. Waste Conf.*, Purdue University, Lafayette, Ind., 1958,

18. C. N. Sawyer, "Experiences in Aerated Lagoon Design and Operation," in *Advances in Water Quality Improvement*, Vol. I, University of Texas Press, Austin, Texas, 1966.

19. J. Mancini and E. Barnhart, "Industrial Waste Treatment in Aerated Lagoons," Advances in Water Quality Improvement, Vol. 1, University of Texas Press, Austin, Texas, April 1966.

20. A. Downing, in *Advances in Water Quality Improvement*, Vol. I, University of Texas Press, Austin, Texas, 1966.

21. K. Wuhrmann, "Nitrogen Removal in Sewage Treatment Processes," *Verhandl. Intern. Verein, Limnol.*, **XV**, 580–596 (1964).

22. A. L. Downing, "Population Dynamics in Biological Systems," *Proc. 3rd Intern. Conf. Water Pollution Research,* Munich, 1966, Water Pollution Control Federation, Washington, D. C.

23. R. E. Speece and P. L. McCarty, "Nutrient Requirements and Biological Solids Accumulation in Anaerobic Digestion," in *Advances in Water Pollution Research*, Vol. 2, Pergamon Press, Oxford, 1964.

24. W. N. Torpey, "Loading to Failure of a Pilot High Rate Digester," *Sewage Ind. Wastes*, **27**, 121 (1955).

25. A. Lawrence and P. L. McCarty, *Kinetics of Methane Fermentation in Anaerobic Waste Treatment*, Tech. Rep. 75, Department of Civil Engineering, Stanford University, Stanford, Calif., Feb. 1967.

26. S. Balakrishnan et al., "Organics Removal by a Selected Trickling Filter Media," *Water Wastes Eng.*, **23** (1969).

27. S. Balakrishnan, Ph.D. Thesis, University of Texas, Austin, Texas, 1968.

28. E. F. Gloyna, "Basis for Waste Stabilization Pond Design," in *Advances in Water Quality Improvement*, Vol. I, University of Texas Press, Austin, Texas, 1966.

29. C. P. Fischer et al., in *Advances in Water Quality Improvement*, Vol. I, University of Texas Press, Austin, Texas, 1966.

30. A. Estrada, "Cost and Performance of Sludge Digestion Systems," *Proc. ASCE*, **86**, SA3, 111 (1960).

31. C. N. Sawyer, "An Evaluation of High Rate Digestion," in *Biological Treatment of Sewage and Industrial Wastes*, Vol. II, J. McCabe and W. W. Eckenfelder, eds., Reinhold, New York, 1958.

32. G. J. Schroepfer et al., *Advances in Water Pollution Research*, Vol. 1, Pergamon Press, Oxford, 1964.

33. G. Culp and A. Slechta, "Nitrogen Removal from Sewage," Final Progress Report, U.S.P.H.S. Demonstration Grant 86-01, Feb. 1966.

34. A. L. Downing et al., "Nitrification in the Activated Sludge Process," *J. Inst. Sewage Purification*, (2) 130-158 (1964).

35. H. Heukelekian et al., "Factors Affecting the Quantity of Sludge Production in the Activated Sludge Process," *Sewage Ind. Wastes*, **23**, 945 (1951).

36. W. W. Eckenfelder, *Industrial Water Pollution Control*, McGraw-Hill, New York, 1966.

10

Tertiary Treatment

Tertiary treatment may be categorically defined as treatment for the removal of pollutants not removed by conventional biological treatment processes (activated sludge, trickling filters, aerated lagoons, etc.). These pollutants will include suspended solids, BOD (usually less than 10 to 15 mg/l), refractory organics (usually reported as COD or TOC), nutrients (nitrogen and phosphorus), and inorganic salts.

The characteristics of secondary effluents will vary widely, particularly in the case of industrial wastes, and will be to a great degree a function of the characteristics of the untreated waste. The general characteristics of secondary effluents from sewage and industrial waste secondary treatment plants are summarized in Table 10.1. Specific effluent characteristics are shown in Table 10.2.

Prior to considering tertiary treatment needs, it is first necessary to establish water-quality requirements for specific water uses. In the United States today, increasing emphasis is being placed on the removal of phosphorus and the removal of unoxidized nitrogen which will exhibit a long-term oxygen demand in the receiving waters. For example, a recent State of Michigan requirement for a major municipality was removal of 80 percent of the total phosphate and an average and maximum daily 20-day BOD in the effluent of 8 mg/l and 15 mg/l, respectively, during the summer months. The average suspended solids must not exceed 10 mg/l.

It is probable that requirements of this type will become prevalent in many urbanized parts of the United States and in other parts of the world in the near future. More stringent requirements relating to refractory organics and total inorganic

214

**Table 10.1. General Characteristics of
Secondary-Treatment Effluents**

Secondary Effluent	Concentration
Soluble BOD	< 10 mg/l
COD[a]	COD_I - $BOD_{u/0.9}$ + COD_B (mg/l)
Suspended solids	< 20 mg/l activated sludge; > 50 mg/l aerated lagoon
Nitrogen[b]	N_I - $0.12 \Delta X_v$ (lb)(kg)
Phosphorus[c]	P_I - $0.023 \Delta X_v$ (lb)(kg)
Total dissolved solids (inorganic)	No appreciable change with treatment

[a] The effluent COD may be increased by 1 to 3 percent of the influent COD, owing to the production of nonbiodegradable metabolic products.

[b] Total nitrogen: NH_3—N, NO_3—N, and organic; ΔX_v refers to the biological sludge wasted; it does not apply to aerated lagoons.

[c] Phosphorus may also be precipitated in the process, depending on the pH and the calcium content.

solids will become necessary in more arid areas, where extensive water reuse is necessary for industrial expansion.

In recent years, laboratory and pilot-plant research (and in some cases, large demonstration plants) has established various combinations of tertiary treatment to meet varying effluent quality requirements. The unit processes that have been investigated for specific treatment requirements are tabulated in Table 10.3. The various applicable processes are shown as a substitution diagram in Fig 10.1. A process employed in South Africa is shown in Fig 10.2. Three process combinations which have been used and their effluent characteristics are shown in Fig. 10.3–10.5. Other process combinations could and have been employed to meet specific quality levels.

To place tertiary treatment in perspective, a more detailed discussion of the principal processes now available for the removal of specific pollutants is presented next.

Table 10.2. Characteristics of Secondary Effluents

Constituent, mg/l	Sewage, Stevenage, G.B.	Sewage, Amarillo, Tex.	Refinery[a]	Refinery[b]	Petrochemical[b]
Total solids	728	557	2900	3000	—
Suspended solids	15	11	14	17	—
Volatile suspended solids	—	—	10	10	11
BOD	9	10	2	4	132
COD	63	—	99	112	—
Phosphate, as P	9.6	9.0	—	—	—
Nitrogen, as N	43.9	22.3	—	—	—
Chlorides	69	83	—	1640	—
pH	7.6	7.7	6.8	6.6	7.9

[a] Activated sludge.
[b] Extended aeration.

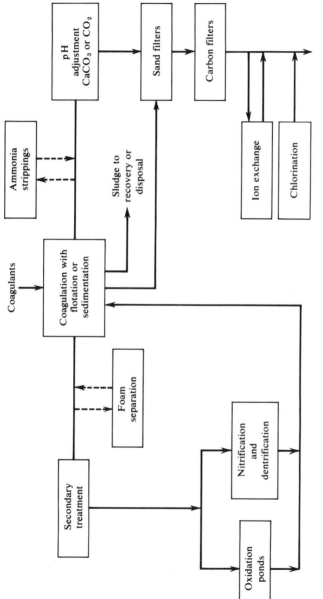

Figure 10.1 General process for water renovation.

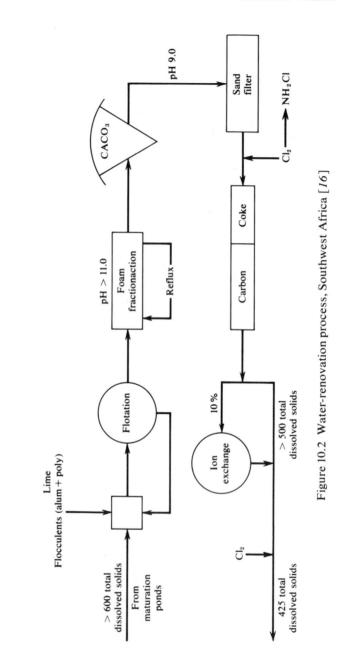

Figure 10.2 Water-renovation process, Southwest Africa [16]

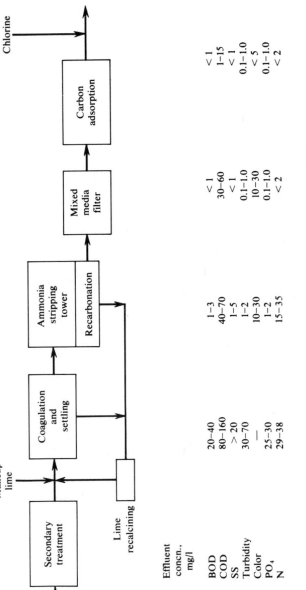

Secondary treatment	Coagulation and settling	Ammonia stripping tower / Recarbonation	Mixed media filter	Carbon adsorption
Effluent concn., mg/l				
BOD	20–40	1–3	< 1	< 1
COD	80–160	40–70	30–60	1–15
SS	> 20	1–5	< 1	< 1
Turbidity	30–70	1–2	0.1–1.0	0.1–1.0
Color	—	10–30	10–30	< 5
PO₄	25–30	1–2	0.1–1.0	0.1–1.0
N	29–38	15–35	< 2	< 2

Figure 10.3 Tertiary treatment processes, Lake Tahoe. [9]

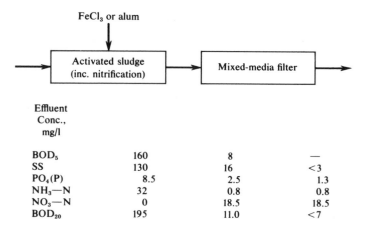

Figure 10.4 Tertiary-treatment process for phosphorus removal.

NUTRIENT REMOVAL

Phosphorus Phosphorus is recognized today as a primary element responsible for eutrophication in natural bodies of water. Several methods are presently available at varying costs and efficiencies for phosphorus removal.

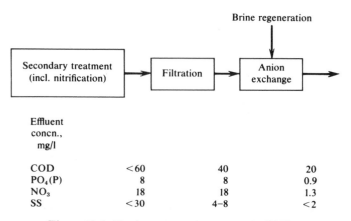

Figure 10.5 Tertiary treatment processes [11]

Table 10.3. Tertiary-Treatment Processes
for the Removal of Specific Pollutants

Unit Process	Major Elements Removed	Additional Features
Filtration—sand, diatomite, or mixed media	Suspended solids	Removal of BOD, COD, PO_4 in suspended form
Filtration plus coagulation—mixed media	Suspended solids and phosphate, color and turbidity	As above plus colloidal solids
Coagulation	Color, turbidity, and PO_4	Some COD and BOD removed
Air stripping	NH_3	High pH required
Nitrification and denitrification	Nitrogen	Ultimate BOD reduced
Carbon adsorption	COD or TOC	Reduction in color and residual suspended solids
Ion exchange	PO_4, nitrogen, total dissolved solids	Resins selected for specific purposes
Reverse osmosis	Organics and inorganics	Pretreatment required to avoid membrane fouling; treatment or disposal of residue
Electrodialysis	Inorganic salts	Pretreatment required to avoid membrane fouling; treatment or disposal of residue

One of the most promising procedures is the precipitation of phosphorus in the activated sludge process. Recently, several investigators [1, 2] have reported biological removal of phosphorus through luxury biological uptake. It has been well established that organic removal resulting in microbial growth results in about 2.6 percent (by weight) phosphorus in the cells grown. In domestic sewage, this would account for 20 to 40 percent removal of phosphorus or 0.8 to 2.0 mg/l in the biological process. Luxury uptake implies removals of 80 to 95 percent, considerably in excess of that required for normal biological growth. As luxury uptake has been neither predictable nor consistent, considerable doubt has been expressed regarding this mechanism. Jenkins [3] has recently explained the luxury-uptake phenomena in terms of chemical precipitation of CaH PO$_4$ as related to the pH and calcium content of the carrier water.

Consistent and predictable phosphate removal has been obtained by the addition of aluminum or iron salts to the aeration tank. Thomas [4] has obtained 92 percent removal of total phosphate on plant scale by the addition of 10 mg/l FeCl$_3$ to activated sludge aeration tanks in Mannedorf, Switzerland.

Recent studies by Barth and Ettinger [5] showed phosphorus removal in excess of 90 percent by the addition of coagulants to the aeration tank. Some of their results and others are summarized in Table 10.4.

Table 10.4. Removal of Phosphorus by
Coagulant Addition to the Aeration Tank

Coagulant Added	Initial Phosphorus(P), mg/l	Percent P removal	Reference
Calcium, 150 mg/l	4–12	64	[5]
Magnesium, 20 mg/l	4–12	50	[5]
[1]Fe^{+++}, 15 mg/l[a]	4–12	75	[5]
[1]Al^{+++}, 20 mg/l[a]	4–12	70	[5]
Al^{+++}, 30 mg/l + 20 mg/l of Ca	4–12	90	[5]
NaAl(OH)$_4$, 10 mg/l	8–15	94	[5]
FeCl$_3$, 15 mg/l	8.5	82	[4]
FeCl$_3$, 10 mg/l	7.9	92	[4]

[a]Turbid effluent due to reduction in pH.

It is significant to note that the chemical addition aided settling and compaction of the activated sludge in secondary clarifiers. A chemical requirement of approximately 1.1 to 2.0 times the theoretical phosphorus requirement is needed to attain effluent qualities with a phosphorus concentration less than 0.5 mg/l. No problems in sludge digestion were observed, although it is desirable to keep sludge supernatant liquor, with its high phosphate content, out of the aeration system. Ferric ions will be reduced to ferrous rendering the phosphate soluble during anaerobic digestion.

Thomas [4] in plant-scale studies showed that the effluent phosphorus from biological treatment was 4.4 mg/l without iron precipitation, as compared to 0.6 mg/l with 10 mg/l of iron added. The value of iron-containing return sludge was demonstrated by the fact that a postprecipitation process as proposed by Wuhrmann [6] consumed 2 to 3 times the iron plus an excess of 100 mg/l of lime to achieve the same effluent quality.

Several processes have been developed for the chemical removal of phosphorus. Phosphorus removal by coagulation occurs by precipitation and/or adsorption on an insoluble solids phase. Lime coagulation primarily results in precipitation, while the mechanism with alum or iron salts appears to be adsorption on the hydrated oxide floc particles. Buzzell and Sawyer [7] have shown that at pH levels of 10 to 11 in the primary sedimentation tanks, BOD removal of 55 to 70 percent, nitrogen removal of 25 percent, coliform removal of 99 percent, and phosphate removal of 80 to 90 percent can be expected. To achieve these pH levels, lime must be added at dosages of 2.5 times the alkalinity (for 100 mg/l alk.) to 1.7 times the alkalinity at 300 mg/l. The resulting sludge is about 1 percent of the volume of wastewater treated. Rand and Nemerow [8] have studied the reclaiming and reusing of lime by recalcination of the lime sludge. The lime is extracted from the ash after slaking. Owing to the organics present, the sludge in many cases may support its own combustion. Because of the presence of inerts in the recalcined sludge, lime makeup of about 36 percent is estimated.

Buzzell and Sawyer [7] showed that the low BOD remaining after lime treatment of primary sewage may render acttivated sludge unsuitable. Secondary treatment by stabilization ponds would prove most satisfactory.

The PEP process utilizes lime addition to the primary clarifier with flocculation and sludge recirculation. The indication is that with sludge recirculation the lime addition can be reduced by as much as 50 percent to achieve the same residual soluble phosphate and that comparable removal could be achieved at approximately 1 pH unit lower than that achieved without recirculation. Clarifier overflow rates of 2000 gpd/ft^2 (81,500 1/day/m^2) are suggested. BOD removals of 60 to 75 percent at pH 9.5 to 10 have been reported.

The Dow process precipitates the phosphorus by the addition of ferrous, ferric, or aluminum salts, some alkalinity, and a high-molecular-weight polyelectrolyte. Typical chemical requirements are 10 to 25 mg/l of $FeCl_2$ (as Fe), 30 to 40 mg/l of alkalinity (as $CaCO_3$), and 0.3 to 0.5 mg/l of polyelectrolyte. Phosphorus removals in excess of 80 percent with high suspended solids and BOD removals are reported. The process can be employed in the primary step or as a tertiary treatment step.

Post precipitation of phosphate has been employed in several cases. Wuhrmann [6] successfully precipitated 90 percent of the total phosphorus in secondary effluent with the addition of 10 to 20 mg/l of Fe^{+++} and the addition of lime to raise the pH to 8.0 to 8.3. Lime dosages of 300 to 350 mg/l (depending on the alkalinity of the water) were used. Phosphate from secondary effluent has been successfully removed at Lake Tahoe by precipitation with lime [9].

Nitrogen   Nitrogen is considered a pollutant both because it exerts an oxygen demand in the unoxidized state (NH_3—N) and because of its contribution to eutrophication of natural waters. Several processes have been developed for the removal of nitrogen both in the oxidized and unoxidized state.

Ammonia nitrogen can be air stripped from solution when it is in the un-ionized form of ammonia. At pH levels of 6 to 8 the ammonia nitrogen is predominately in the ionized NH_4^+ form. At pH levels above 10.0, all the nitrogen is in the ammonia form and removable by aeration. Lime is usually used for pH adjustment. Economics dictate the use of lime recovery in large installations. Even with lime recovery, approximately 40 percent lime makeup is required. The stripping process can be employed either before or after secondary treatment [9].

Ninety-two percent removal of ammonia has been achieved by air stripping through a packed tower of 20-ft depth at pH 10 to 11. An air rate of 300 ft^3/gal (2.24 m^3/l) with a liquid rate of 3.0 gpm/ft^2 (122 1/min/m^2) was used. Increasing the removal efficiency to 98 percent required 800 ft³/gal (5.98 m³/l).

The primary variables in ammonia stripping are pH, air rate, tower depth, and hydraulic loading to the tower. These are shown in Fig. 10-6

Organic nitrogen or oxidized nitrogen (NO_3^-) will not be removed in this process. The removal efficiency drops to 50 to 60 percent during cold-weather operation. Because ammonia stripping is conducted at an alkaline pH, phosphorus removal will also be achieved with significant BOD and suspended-solids removal. If the stripping process is used as a pretreatment prior to biological treatment, enough nitrogen must be left in the effluent to satisfy the nutritional requirements in the biological process. pH adjustment by recarbonation of the effluent will usually be required.

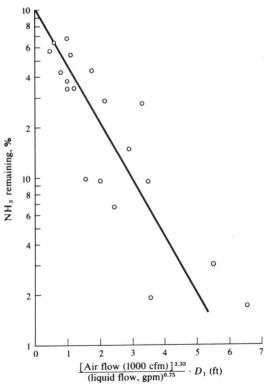

Figure 10.6 Ammonia removal by air stripping. pH 11.5; tower packed with 1 1/2 inch × 3/8 inch redwood slats mounted on 2 inch centers horizontally and 1 1/2 inch centers vertically [9].

Recently an ion-exchange resin, clinoptilolite, which is selective for NH_4^+, has been studied for the removal of NH_4^+ from clarified secondary effluents [10]. The resin columns are regenerated with lime and total recycle of the regenerant used after ammonia removal by air stripping.

Nitrate nitrogen (after oxidation in the activated sludge process) has been removed by anion exchange regenerated with brine by Eliassen and Bennett [11]. This ion-exchange process also removes phosphate and some other ions. Pretreatment by filtration is essential. The resin is restored by treatment with acid and methanol. The performance and cost figures for this process are shown in Table 10.5.

The process which has received the most extensive study is nitrification followed by denitrification [12]. Nitrification is achieved by adjusting the sludge age (or organic loading in the case of domestic sewage) to be greater than the growth rate of *Nitrosomonas*, which oxidizes ammonia to nitrite.

At 20°C, a sludge age of approximately 3.5 days is required, increasing to 5.5 days at 15°C [12]. This is equivalent to an organic loading of approximately 0.3 lb of BOD_5/day/lb of MLSS. Nitrification requires 4.6 lb of O_2/lb of N oxidized at dissolved oxygen levels in excess of 2.0 mg/l.

The increase in aeration tank volume and horsepower relative to a conventional design loading of 0.45 lb of BOD_5/day/lb of MLSS, a BOD_5 of 300 mg/l, and an MLSS of 3500 mg/l is shown in Table 10.6.

Denitrification results from the action of facultative organisms in the absence of molecular oxygen (actually some denitrification will occur in the presence of oxygen under certain conditions). Since the rate of denitrification is directly proportional to the respiration rate or the organic loading rate, long detention periods are required in the absence of organic substrates.

Several process alternatives have been proposed for denitrification. Wuhrmann [13] has proposed a mixed basin without aeration following the activated sludge unit in which denitrification occurs through endogenous respiration. Bringmann et al. [14] bypass a portion of the sewage to the denitrification basin to increase the respiration rate. Other processes have bypassed a nitrogen-deficient industrial waste to

Table 10.5. Reported Costs of Tertiary Treatment Processes

Process	Liquid Treated	Plant Size, Mgd (m³/day)	Inf., mg/l		Eff., mg/l	Cost ¢/1000 gal (¢/1000 l) operating	total	Remarks	Reference
Adsorption	Sec. eff.	10 (37,850)	COD_T	47	9.5	4.2 (1.1)	8.3 (2.2)	Includes regeneration	[22]
			COD_D	31	7				
			TOC	13	2.5				
			SS	10	<2				
			Color	30	3				
			NO_3	6.7	3.7				
Ion exchange	Sec. eff.	3 (11,350)	P	8	0.9	15 (4.0)	18 (4.8)	Includes resin restoration with acid and methanol	[11]
			NO_3	18	1.3				
			COD	40	20				
			SS	>15	<2				
Nitrification and denitrification	Act. sludge	5 (18,900)	—		—	—	3.2 (0.8)		[9]
Chemical treatment[a]	Sec. eff.	7.5 (28,400)	—		—	5.2 (1.4)	6.3 (1.7)		[9]
NH_3 stripping	Sec. eff.	7.5 (28,400)	—		—	0.9 (0.2)	1.4 (0.4)		[9]
Separation beds	Sec. eff.	7.5 (28,400)	—		—	2.3 (0.6)	4.0 (1.1)		[9]
Adsorption	Sec. eff.	7.5 (28,400)	—		—	1.6 (0.4)	4.0 (1.1)		[9]
Electrodialysis	Sec. eff.	10.0 (37,850)	—		—	16.0 (4.2)	—	40% TDS removal; pretreatment by filtration or coagulation and residue disposal not included	[9]

[a] See Fig. 10.3 for results.

**Table 10.6. Aeration–Tank Volume
and Horsepower**

Temp., °C	Increase in Aeration Volume, %	Increase in Aeration hp, %
20	0	100
15	55	100
10	135	100

the denitrification tank to increase the respiration rate in the process.

At 12°C and an endogenous respiration level of 5.5 mg of O_2/g of MLSS/hr, a retention period of 4.5 hr is required to remove 30 mg/l of N with an MLSS of 3500 mg/l. At 18°C, with an endogenous respiration level of 7.0 mg of O_2/g of MLSS/hr, a retention period of 3.4 hr is required. The relationships between nitrate removal rate and respiration rate are given on page 173.

Nitrogen removal by denitrification may be variable, owing to load variation in the plant resulting in a variable respiration rate in the denitrification step. Denitrification can also be achieved by filtration through carbon or sand filters. Methanol has been supplied as a supplemental food at a hydraulic loading of 7 gpm/ft² (285 l/min/m²) in a 10-ft (3.05-m) column [15]. Nitrate removals to 1 mg/l or less have been obtained with the addition of 2 mg/l of methanol per mg/l of NO_3—N.

Stabilization ponds have been used for nutrient removal by algae. The problems associated with this process are the seasonal instability of algal removal of nutrients and the difficulty of separating algae from the treated wastewater. Cillie et al. [16] obtained 63 percent removal of total nitrogen from a Biofilter effluent after 14 days in a pond to which 10 percent settled sewage was added (to aid denitrification). The algae was separated by flotation with alum. Parkhurst [17] showed 50 percent reduction in total nitrogen in a stabilization pond treating settled sewage with a retention period of 60 days. Ninety percent algal removal and substantially complete removal of phosphate was effected by flotation or sedimentation with alum.

SUSPENDED SOLIDS REMOVAL

Filtration is employed for the removal of finely divided suspended material carrying over from secondary clarification or chemical precipitation units. Since PO_4, COD, and BOD are present in suspended form, removal of these constituents will also result from filtration.

Filtration can be accomplished by the use of (1) a microstrainer, (2) diatomaceous earth filtration, (3) sand filtration, or (4) mixed-media filtration.

The microstrainer is a screen in the form of a partially submerged rotating drum or cylinder. Water flows continuously by gravity through the submerged portion from inside the drum to a clear-water storage chamber outside the drum. Cleaning is carried out by backwashing with sprays of product water. Backwash requirements are less than 5 percent of the feed volume. At low suspended-solids levels, loading rates of 4 to 7 gpm/ft² (163 to 285/min/m²) are employed. Reported removals of suspended solids, BOD, and turbidity from secondary effluent are 73 to 89, 61 to 81, and 60 to 76 percent, respectively [18]. Studies at Chicago showed 71 and 74 percent removal of suspended solids and BOD, respectively.

Diatomaceous earth filtration is a mechanical separation that employs a layer of filter aid (diatomeaceous earth built up on a septrum to screen out suspended solids). As filtration proceeds, deposited solids will build up on the precoat, resulting in an increase in pressure drop. The filter run can be increased by the addition of filter aid to the effluent (body feed), which maintains the porosity of the cake. When the pressure drop becomes too great to continue filtration, the filter is backwashed and a new precoat applied. The minimum particle size which can be removed is related to the particle size of the filter aid (about 0.5 to 1.0 μ). Flow rates of 0.5 to 1.0 gpm/ft² (20.4 to 40.7 l/min/m²) result in turbidity removals in excess of 80 percent [18].

Sand filtration is usually employed following chemical coagulation and preceding carbon adsorption or ion exchange. The average filter run before backwashing is related to the solids loading to the filter.

Principal problems result from varying suspended solids content of the wastewater, causing blinding and shortened filter runs.

The application of mixed-media filtration offers improvement in this regard [19]. In this unit, three or more materials of different specific gravities are used, increasing from 1.6 at top to over 4 at the bottom as shown in Fig. 10.7.

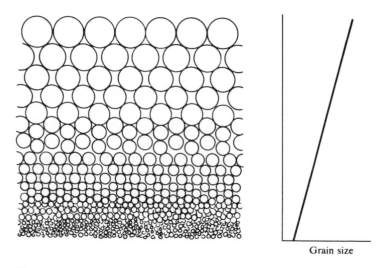

Grain size

Figure 10.7 Cross section through ideal filter uniformly graded from coarse to fine from top to bottom.

Filter depths of 24 to 30 in. (61 to 76 cm) flow rates of 5 to 10 gpm/ft² (204 to 407 l/min/m²) and backwash rates of 15 to 20 gpm/ft² (610 to 815 l/min/m²) are normally used in secondary effluent filtration. Low-effluent suspended solids (<3 mg/l) are readily obtainable [19].

A modification of the mixed-media filter is the application of coagulated effluent directly to the filter. In some cases of high coagulant doses, a primary filter consisting entirely of coarse coal is used.

Chemical coagulation is employed for the initial treatment of secondary effluents. Coagulation is accomplished using alum and a polyelectrolyte or lime. If the effluent is from a postoxidation pond, alum (100 to 250 mg/l) will effectively flocculate the algae and other solids in the effluent. Separation is accomplished by flotation or sedimentation. Typical results obtained are shown in Table 10.7.

ORGANIC REMOVAL

Adsorption   There is an increasing discharge to sewers of exotic organics, detergents, pesticides, and other substances that are resistant to biological oxidation. The removal of these substances is essential in many cases to render the water suitable for reuse. The method finding greatest application today is adsorption on activated carbon.

Adsorption can be accomplished with powdered carbon, which is mixed, flocculated, and settled from the waste or with granular carbon of 40 to 80 mesh in columns or counterflow fluidized beds. Because of operational efficiency and economics the later methods have found the greatest application.

It is generally desirable to provide filtration of the water for removal of finely divided suspended matter prior to application to a carbon bed. It has been found that organic breakthrough is associated with suspended solids in unclarified wastewater [21]. Removal of organics in a carbon bed occurs by adsorption of the less polar molecules, filtration of large particles, and partial deposition of colloidal material. The removal obtained depends primarily on the contact time between the carbon and the water. Because amino acids and sodium salts of fatty acids with less than eight carbon atoms and glucose are only slightly adsorbed, high-efficiency adsorption implies efficient secondary treatment.

As the water passes through the bed the carbon nearest the feed point eventually becomes saturated and must be replaced with fresh or reactivated carbon. A countercurrent flow using multiple columns in series is the most efficient. The first col-

Table 10.7. **Chemical Coagulation of Sewage Effluent**

Characteristic	Stevenage[a]		Windhoek[b]		Lancaster[c]		Tahoe[d]	
	Inf.	Eff.	Inf.	Eff.	Inf.	Eff.	Inf.	Eff.
BOD	3	1	30	1	25	4	20–40	<1
COD	41	25	12[e]	1	170	55	80–160	30–60
SAA	0.8	0.7	8	4	2.5	1.5	1.1–2.9	1.1–2.9
TOC	11	10	—	—	—	—	—	10–18
P	8	1.5	10	Nil	13	0.1	8–10	<0.3
N	1.2	0.8	35	15	13	5.3	—	—
SO$_4$	66	99	108	220	55	220	—	—
Cl	74	77	—	—	—	—	—	—
pH	7.8	7.4	8.5	8.0	8.5	6.6	—	—
SS	10	3	—	—	70	14	5–20	0.2–3.0
TDS	—	—	—	—	585	610	—	—

[a] Coagulation with alum, 40 mg/l; from Eden et al. [20].
[b] Oxidation pond effluent; 220 mg/l of alum, followed by lime and chlorine; from Cillie et al. [16].
[c] Coagulation of oxidation pond effluent with 300 mg/l of alum followed by dissolved air flotation; from Parkhurst [15].
[d] From Culp [9].
[e] Permanganate value.

umn is replaced when exhausted and the flow of water changed to make that column the last in the series.

In the adsorption columns at Lake Tahoe [9] the clarified separation-bed effluent flows upward through the carbon and the spent carbon is removed from the bottom of the filter. The average detention time through the carbon is 24 min.

In large installations, economy necessitates the regeneration and reuse of the carbon. At the Lake Tahoe plant, this is accomplished by dewatering for 1 hr to 43 percent moisture followed by thermal regeneration at 1650 to 1680 °F (898.9 to 910°C). Steam is injected at a rate of 1 lb (kg) of steam/lb (kg) of carbon. An estimated carbon loss of 5 to 10 percent can be expected from each regeneration.

The removal efficiency of organics from wastewaters as developed from adsorption isotherms is summarized in Table 10.8.

An extensive pilot-plant study treating secondary effluent has been reported by Parkhurst el at. [22]. Their results showed a 78 percent removal of soluble COD to a residual of 7 mg/l at a removal rate of 55 to 60 lb (kg) of COD/100 lb (kg) of carbon. Operation was with a four-stage packed-bed downflow granular activated carbon column. The data from this operation are summarized in Table 10.5.

Adsorption can also be accomplished using powdered activated carbon. The small particle size renders it unsuitable for column operation. Conventional practice involves a combination of mixing, flocculation, settling, and, in some cases, filtration. Countercurrent operation using multiple units may also be employed. Reaction is essential for economic operation. Dosages of 100 to 200 mg/l of carbon with 0.5 to 1.0 mg/l of polyelectrolyte to aid flocculation have resulted in reductions of TOC from 11 to 36 mg/l to 0.9 to 2.0 mg/l.

INORGANIC REMOVAL

Ion Exchange The removal of 40 percent of the inorganic dissolved solids is essential if a water is to receive multiple industrial or municipal reuse. This represents the increase for each reuse cycle.

Table 10.8. Adsorption Characteristics of Secondary Effluents as Measured by Isotherm Tests

Waste	Pretreatment	Initial concn., C_0, mg/l COD or TOC	lb (kg) COD or TOC removed/100 lb (kg) of carbon, at C_0	Minimum residual concn., mg/l	Reference
Sewage	None	118[a]	76	10	[27]
Sewage	None	23.8[a]	17	8	[27]
Sewage	None	55[a]	22	25	[27]
Sewage	None	40[a]	25	8	[28]
Refinery	None	23[b]	59	5	[29]
Refinery	None	16[b]	48	4.6	[29]
Petrochemical	None	33[b]	36	8.8	[29]
Sewage[c]	None	31[a]	—	7.0	
		13[b]	—	2.5	

[a] Measured as COD.
[b] Measured as TOC.
[c] Continuous carbon columns.

Sanks and Kaufmann [23] used strong-acid and weak-base resins in fixed beds with removal of carbon dioxide between the cationic and anionic units. An exchange efficiency in the cation unit of 93 percent at a regeneration level of 1.25 lb of H_2SO_4/ft^3 (20 kg/m^3) (5 percent acid) was obtained with regenerant recovery. The overall efficiency (including regenerant makeup and reuse water was 84 percent, with 20 to 25 percent leakage). The anion unit was regenerated with 1 lb of NH_3/ft^3 (16 kg/m^3) (2 percent) and resulted in 86 percent exchange efficiency with 35 percent leakage. Rinse-water requirements ranged from 11 to 16 gal/ft^3 (1470 to 2140 l/m^3). Spent regenerant was 3.3 percent of the water treated. Studies have also been conducted using weakly basic phenolic resins, for reversible sorption of ABS and other organics.

The recently developed Desal [24] process uses a weakly basic anion exchange resin with a favorable chloride–bicarbonate selectivity coefficient to form a resin bicarbonate salt when reacted with dissolved carbon dioxide in wastewater streams (see Fig. 10.8). A weakly acid cationic resin in the

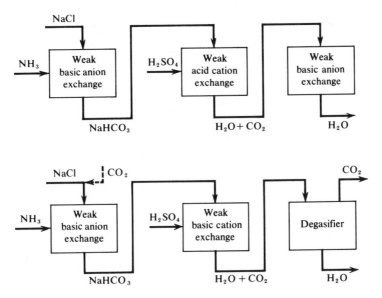

Figure 10.8 Desal ion-exchange process.

hydrogen form is used in conjunction with the anionic resin. Several alternative operations have been proposed. In one alternative, effluent is passed through an anion unit, exchanging anions for bicarbonate. In the second unit (cation) the bicarbonate salts are converted to carbonic acid. A second anion unit receives the flow from the cation unit, removing the carbonate as bicarbonate exhaustion; the first unit is regenerated back to the free base form with ammonia, caustic, or lime and the cation unit regenerated with sulfuric acid. The third unit is already in the bicarbonate form, so the flow pattern is reversed for the next cycle. A second alternative operation replaces the third unit with a degasifier for CO_2 to the first unit.

Electrodialysis Electrodialysis involves removing the inorganic ions from water by impressing an electrical potential across the water, resulting in the migration of cations and anions to the cathode and anode, respectively. By alternately placing anionic and cationic permeable membranes a series of concentrating and diluting compartments result. The amount of electric current required for the demineralization is proportional to the number of ions removed from the diluting compartments. An electrodialysis cell is shown in Fig. 10.9. To minimize membrane fouling and maintain long runs, it is essential that turbidity suspended solids, colloids, and trace organics be removed prior to the electrodialysis cell. Large organic ions and colloids are attracted to the membranes by electrical forces and result in fouling of the membranes. This results in an increase in electrical resistance which at constant stack voltage decreases the current and hence the demineralizing capacity of the equipment. Turbidity in the feed has a major effect on membrane fouling. Conventional methods of coagulation, filtration, and adsorption would be employed. Brunner [25] obtained 40 percent total dissolved solids removal with a concentrate of 10 percent (by volume) of the influent water. The estimated operating costs are 12 to 16 cents/1000 gal (3.2 to 4.2 cents/1000 l) for a 10-Mgd (37,850 m³/day) plant, not including pretreatment or concentrate disposal.

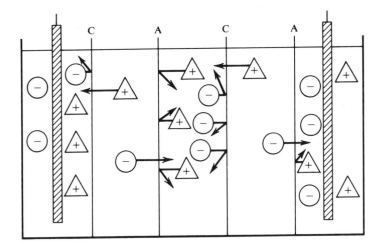

Figure 10.9 Electrodialysis Cell (△—Cation; ⊖—Anion; A—Anion-permeable membrane; C—Cation-permeable membrane)

Reverse Osmosis   Reverse osmosis can be employed for the removal or concentration and recovery of ions such as chromate and phosphates and for the selective rejection of organic compounds. In reverse osmosis, water containing dissolved materials is placed in contact with a suitable membrane at a pressure in excess of the osmotic pressure of the solution, as shown in Fig. 10.10. Under these conditions water with a small amount of dissolved material permeates the membrane and the dissolved materials are concentrated. The membranes are a specially prepared cellulose acetate. High concentrations of soluble material can cause a scale, and suspended solids can form a fouling layer. This is reduced by moving water past the membrane surface at a velocity sufficient to cause turbulence. The area of transfer is significant. DuPont has employed hollow fibers in a size range of 25 to 250 μ as the membrane media [26]. Treatment units 12 in. by 7 ft (30.5 cm by 2.13 m) will contain 15 to 80 million hollow fibers with surface areas of 50,000 to 80,000 ft²(4650 to

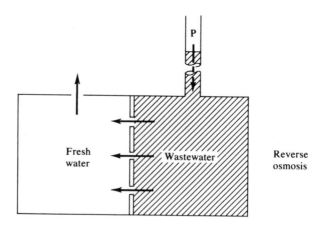

Figure 10.10 Schematic presentation of reverse osmosis.

7440 m^2). Results on one organic waste reduced the BOD and COD from 108 and 1035 mg/l to 35 and 100 mg/l, respectively, at 600 psig (42 kg/cm^2).

Acidification of the feed may be necessary to prevent calcium carbonate scale formation. TDS removals of 95 percent with substantial removals of COD, phosphate, and ammonia have been reported.

REFERENCES

1. D. Vacker et al., "Phosphate Removal through Municipal Wastewater Treatment at San Antonio, Texas," *J. Water Pollution Control Federation*, **39**, No. 5, 750 (1967).
2. G. V. Levin, and J. Shapiro, "Metabolic Uptake of Phosphorus by Wastewater Organisms," *J. Water Pollution Control Federation*, **37**, No. 6, 800 (1965).
3. D. Jenkins, private communication.
4. E. A. Thomas, "Phosphate Precipitation in the Uster Treatment Plant and Removal of Iron Phosphate Sludge," *Viertel. Naturforsch. Ges. Zurich* 111, 309-318, (1966).
5. E. F. Barth and M. Ettinger, "Mineral Controlled

Phosphorus Removal in the Activated Sludge Process," *J. Water Pollution Control Federation,* **39**, No. 8, 1362 (1967).

6. K. Wuhrmann, "Nitrogen and Phosphorus Removal," *Schweiz. Z. Hydrol.* **16**, 520-558 (1964).

7. J. C. Buzzell and C. N. Sawyer, "Removal of Algal Nutrients from Raw Wastewater with Lime," *J. Water Pollution Control Federation,* **39**, No. 10, R16 (1967).

8. M. C. Rand and N. L. Nemerow, *Removal of Algal Nutrients from Domestic Wastewater,* prepared for N.Y. State Health Department, 1965.

9. A. F. Slechta, and G. L. Culp, "Water Reclamation Studies at the South Tahoe Utility District," *J. Water Pollution Control Federation,* **39**, No. 5, 787 (1967).

10. B. W. Mercer, et al., "Ammonia Removal from Wastewater by Selective Ion Exchange," Presented to Am. Chem. Soc., April 5, 1968, San Francisco, Calif.

11. K. Eliassen, and G. Bennett, "Anion Exchange and Filtration Techniques for Wastewater Renovation," *J. Water Pollution Control Federation,* **39**, No. 10, R81 (1967).

12. W. W. Eckenfelder, "Water Reuse," *Symp. Ser., A.I.Ch.E.,* **63** (1967).

13. K. Wuhrmann, "Nitrogen Removal in Sewage Treatment Processes," *Intern. Congr. Terminology,* Madison, Wisc., 1962.

14. G. Bringmann et al., "Model Experiments on the Biological Removal of Nitrogen from Treated Sewage," *Gesundh. Ingr.,* **80**, 364-367 (1959).

15. S. Balakrishnan, *Kinetics of Biochemical Nitrification and Denitrification,* Ph.D. Thesis, University of Texas, Austin, Texas, 1968.

16. G. G. Cillie et al., "The Reclamation of Sewage Effluents for Domestic Use," in *Advances in Water Pollution Research,* Vol. 2, Water Pollution Control Federation, Washington, D.C., 1967.

17. J. D. Parkhurst, Discussion in *Advances in Water Pollution Research,* Vol 2, Water Pollution Control Federation, Washington, D.C., 1967.

18. Summary Report — Advanced Waste Treatment, July 1964 - July 1967, U.S. Department of Interior, FWPCA, Washington, D.C., Publ. WP-20-AWTR-19.

19. G. Culp, "High Rate Clarification of Wastewaters," *Proc. Sanit. Eng. Conf.* University of Kansas, Lawrence, Kansas, 1968.

20. Water Pollution Research Branch, Her Majesty's Printing Office, London, 1967.

21. D. F. Bishop et al., "Studies on Activated Carbon Treatment," *J. Water Pollution Control Federation,* 39, No. 2, 188 (1967).

22. J. D. Parkhurst et al., "Pomona Activated Carbon Plant," *J. Water Pollution Control Federation,* 39, No. 10, R69, (1967).

23. R. L. Sanks, and W. Kaufmann, "Partial Demineralization of Brackish Waters by Ion Exchange," *J. Sanit. Eng. Div., ASCE,* 92, No. SA6 (1966).

24. D. G. Downing et al., "Desal Process — Economic Ion Exchange System," presented at the A.I.Ch.E. Meeting, Salt Lake City, Utah, May 1967.

25. C. A. Brunner, "Pilot Plant Experiences in Demineralization of Secondary Effluent Using Electrodialysis," *J. Water Pollution Control Federation,* 39, No. 10, R1 (1967).

26. W. T. Robinson and R. J. Maltson, "A New Product Concept for Improvement of Industrial Water," Presented at Annual Meeting, Water Pollution Control Federation, New York, 1967.

27. R. S. Joyce, and V. A. Sukenik, *Feasibility of Granular Activated Carbon Adsorption for Wastewater Renovation,* AWTR-15, U.S. Department of Health, Education, and Welfare, Washington, D.C., 1965.

28. D. S. Davis, and R. A. Kaplan, "Removal of Refractory Organics from Wastewater with Powdered Activated Carbon," *J. Water Pollution Control Federation,* 38, 442-450 (1966).

29. N. K. Burleson, *Carbon Adsorption in Industrial Water Pollution Control,* M.S. Thesis, University of Texas, Austin, Texas 1968.

11

Sludge Handling and Disposal

Most of the treatment processes normally employed in water-pollution control yield a sludge from a solids–liquid separation process (sedimentation, flotation, etc.) or produce a sludge as a result of a chemical (coagulation) or a biological reaction. These solids will undergo a series of treatment steps involving thickening, dewatering, and final disposal. Organic sludges may also undergo treatment for reduction of the organic or volatile content prior to final disposal.

The increase in solids content which might be expected through the treatment sequence is shown in Fig. 11.1. In gen-

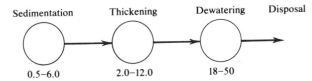

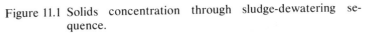

Figure 11.1 Solids concentration through sludge-dewatering sequence.

eral, gelatinous-type sludges such as alum or activated sludge will yield the lower concentrations, whereas primary and inorganic sludges will yield the higher concentrations in each process sequence.

Figure 11.2 shows a substitution flow sheet for the various alternative processes available for sludge dewatering and disposal. The processes selected will depend primarily on the

241

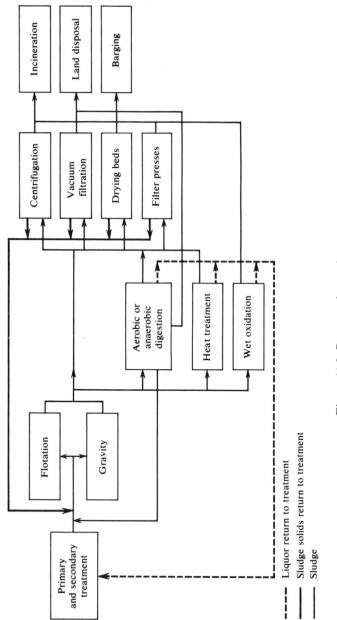

Figure 11.2 Process alternatives.

- - - - Liquor return to treatment
———— Sludge solids return to treatment
———— Sludge

nature and characteristics of the sludge and on the final disposal method employed. For example, activated sludge is considerably more effectively concentrated by flotation than by gravity thickening. Final disposal by incineration requires a solids content that will support its own combustion. In some cases the process sequence is apparent from experience with similar sludges or by geographical or economic constraints. In other cases an experimental program must be developed to determine the most economical solution to a particular problem.

SCREENING, DEGRITTING, AND SKIMMING

Although not normally considered as part of the sludge disposal sequence, grit, screenings, and skimmings from domestic sewage treatment require disposal. Screenings are materials present in the raw sewage that are collected on $\frac{1}{2}$ to 2-in. (1- to 5-cm) screens. These materials are usually buried, but in some cases are incinerated or ground into small particles to be disposed of with the primary sludge. The characteristics and disposal of sewage screenings are summarized in Table 11.1.

Grit is usually removed from sewage as one of the preliminary steps prior to primary and secondary treatment. The heavier grit particles are removed by selective deposition though velocity-control chambers (either hydraulic or mechanical). Grit is disposed of by burial, or if washed, by land fill.

Skimmings are scum, grease, and other floatables removed from primary sedimentation tanks. Mechanical or manual skimming facilities may be provided. Depending on the size of the installation, one of several methods of disposal can be employed. These are summarized in Table 11.1.

THICKENING

The first step in most sludge-disposal processes is thickening, either by gravity or by dissolved air flotation. Thickening is advantageous because it (1) improves digester operation and reduces capital cost where this process is to be employed,

Table 11.1. Characteristics and Disposal of Sewage Screenings, Grit, and Skimmings[a]

Material	Quantity, ft³/Mg (m³/m³)	Moisture Content, %	Organic Content, %	Thermal Value, Btu/lb Kcal/Kg	Disposal
Screenings	0.5–6.0[b] (0.004–0.045)	85–95	50–80	1400–3500 (775–1940)	Burial, incineration,[c] grinding, and return to sewage
Grit	1–12 (0.0075–0.090)	14–34	50[d]	—	Grit chambers hydroclones on primary sludge; burial, drying beds, incineration
Skimmings	0.1–7.0[e] (0.008–0.053)	60–90	90–95	8000–18,000 (4440–9960)	Burial, pumped to digesters, dewatering incineration[f]

[a] From Burd [1].
[b] Screen openings of $\frac{1}{2}$ to 2 in. (1.27 to 5.08 cm).
[c] Separate unit, skimmings, refuse, or dewatered sludge incinerator; reduce moisture content to 15%.
[d] Grit washing can reduce organics to 15%.
[e] Depends on industrial waste discharges to sewerage system.
[f] Separate incineration or combined with sludge, provision for high-temperature burning.

(2) reduces sludge volume for land or sea disposal, and (3) increases economy of sludge-dewatering systems (centrifuges, vacuum filters, etc.).

GRAVITY THICKENING

Gravity thickening is accomplished in a tank equipped with a slowly rotating rake mechanism which breaks the bridge between sludge particles, thereby increasing settling and compaction. A typical gravity thickener is shown in Figure 11.3.

Figure 11.3 Gravity-thickening unit. (Courtesy of Eimco Corporation.)

Thickener Design   The primary objective of a thickener is to provide a concentrated sludge underflow. The area of thickener required for a specified underflow concentration is related to the mass loading (lb/ft²/day, kg/m²/day) or to the unit area (ft²/lb/day, m²/kg/day).

The mass loading can be computed from a stirred laboratory cylinder test. For municipal sewage, the mass loading might be expected to vary from 4 lb/ft²/day (19.5 kg/m²/day) for waste-activated sludge to 22 lb/ft²/day (107 kg/m²/day for primary sludge [1].

Edde and Eckenfelder[2] have shown that thickener variables can be correlated through the relationship

$$\frac{C_u}{C_o} - 1 = \frac{K}{(ML)^n} \tag{1}$$

where C_u = underflow solids concentration, percent solids

C_o = feed solids concentration, percent solids

ML = mass loading in lb of solid/ft²/day (kg of solids/m²/day)

K, n = constants

In equation (1), n is a function of the rheological properties of the sludge and K is related to the height of the test column and to the concentration of feed sludge. Some typical values of n are shown in Table 11.2.

Table 11.2.   Typical Values of n—Sludge Thickening

Sludge Type	n
Lime-softening sludge	1.80
Pulp and paper (primary and secondary)	0.70
Activated carbon sludge	0.53
Domestic primary and activated	0.25

The procedure for calculating the mass loading and designing a gravity thickener is detailed in the design procedure and example.

Most thickeners will have a minimum detention period of 6 hr and a hydraulic overflow rate of 400 to 800 gal/day/ft² (16 to 33 m³/day/m²). In some cases, the addition of polyelectrolytes will improve clarification. An increase in sludge compaction may not be experienced at the dosages usually employed.

Procedure for the Design of a Gravity-Thickening Unit

Data to be Obtained
1. Solids to be thickened; average daily lb/day (kg/day); variation in quantity; solids concentration.
2. Variation in solids characteristics.

Design Procedure
1. Develop a series of three batch settling tests for the range of influent solids concentrations which might be expected in full-scale operation as indicated in step 2 above and shown in step 1 of Fig. 11.4.
2. The unit area for each interfacial settling velocity is calculated from the relationship

$$\text{UA} = \frac{(1/C_i) - (1/C_u)}{U_i} \qquad (2)$$

and is shown in step 2 of Fig. 11.4. U_i is the settling velocity as determined from the tangent to the curve and C_i the interfacial concentration at the point of tangency. H_i is the height the sludge would occupy if it all were at the interfacial concentration C_i.

3. The maximum unit area for each feed concentration and selected underflow concentrations is determined for several interfacial settling velocities over the compression portion of the settling curve as shown in step 3 of Fig. 11.4. It is necessary to select the desired underflow concentrations. The ultimate compaction of the sludge can reasonably be estimated by extrapolation of the settling curves in step 1 [or by a plot of log $(H - H_\infty)$ vs. time, in which H_∞ is the height occupied by the sludge at infinite time]. The underflow concentration C_u is usually selected as 0.7 to 0.9 C_∞. (It should be noted that $C_0 H_0 = C_\infty H_\infty$.) C_0 and H_0 are the initial solids concentration (percent) and initial height, respectively, of the test sample.

4. Using the maximum unit area values obtained in step 3 for different feed and underflow concentrations of a plot is made of log $(C_u/C_o) - 1$ vs. the log of the mass loading $(1/\text{UA})$ as shown in step 4 of Fig. 11.4. The coefficients K_b

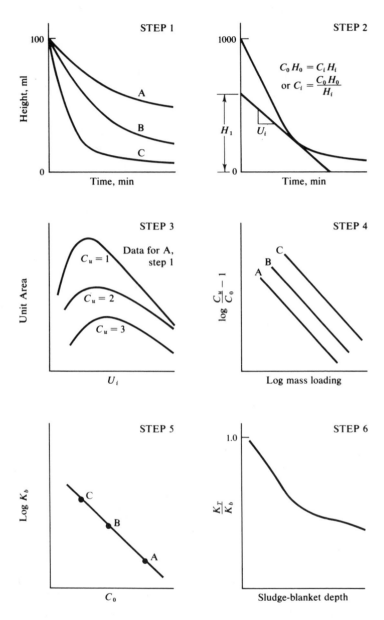

Figure 11.4 Relationships for gravity thickener design

and n are obtained from this plot; n is the slope of the curves and is characteristic of the type of sludge, and K_b is the intercept of the plot and is related to the initial concentration of the suspension and the height of the test cylinder.

5. The intercept, K_b, is related to C_o as shown in step 5 of Fig. 11.4 by plotting log K_b vs. C_o. From this plot, the appropriate coefficient K_b is selected corresponding to the feed concentration C_o.

6. The scaleup factor to a continuous thickener is selected from step 6 of Fig. 11.4, which relates K_T/K_b to the sludge-blanket depth in the continuous thickener.

7. Using the appropriate value for K_T, the allowable mass loading for selected values of feed concentration, and underflow concentration, the mass loading can be computed. The cross-sectional area of the thickener can then be computed.

FLOTATION THICKENING

Thickening through dissolved air flotation is becoming increasingly popular and is particularly applicable to gelatinous sludges such as activated sludge. In flotation thickening, small air bubbles released from solution attach themselves to and become enmeshed in the sludge flocs. The air–solid mixture rises to the surface of the basin, where it concentrates and is removed as shown in Fig. 7.14. The primary variables are recycle ratio, feed solids concentration, air-to-solids ratio, and solids and hydraulic loading rates. Pressures between 40 and 60 psi are commonly employed. Recycle ratio is related to the air-to-solids ratio and the feed solids concentration.

Experience has shown that in some cases dilution of the feed sludge to a lower concentration increases the concentration of the floated solids [3]. Performance data for the thickening of excess activated sludge is shown in Table 11.3. The use of polyelectrolytes will usually increase the solids capture and the thickened sludge concentration.

Table 11.3. Some Results Obtained with Activated Sludge in Dissolved Air Flotation Units

Location	Polymers Used	Feed Suspended Solids Conc., %	Thickened Sludge Sus. Solids Concn., %	Solids Loading, lb/ft²/hr	Overflow rate, gpm/ft²	Reference
Nassau County, N.Y.	No	0.81	4.9	—	—	[17]
Wayne Co., Mich.	No	0.45	4.6	—	—	[17]
Coors Brewery, Golden, Colo.	No	0.77	4.1	—	0.5	[17]
Bernardsville, N.J.	Yes	1.7	4.3	4.3	0.8	[18]
Hatboro, Pa.	Yes	0.7	4.0	2.9	0.4	[18]
Belleville, Ill.	Yes	1.8	5.7	3.8	1.0	[18]
Columbus, Ohio	Yes	0.7	5.0	3.3	1.0	[18]
Fort Worth, Texas	Yes	0.9	4.5	4.5		[18]

CENTRIFUGATION

Centrifugation is enjoying increasing popularity for the dewatering of sewage and waste sludges. While there are several types of centrifuges available, sludge dewatering usually uses a horizontal, cylindrical–conical, solid bowl machine. The centrifuge consists of three major components, a rotating solid bowl that takes the shape of a cylinder and truncated cone section, an inner rotating screw conveyor, and a planetary gear system. A typical unit is shown in Fig. 11.5. Machine variables include bowl speed, pool volume, and conveyor speed [4, 5]. Process variables include the feed rate of solids to the machine, solids characteristics, chemical addition, and temperature.

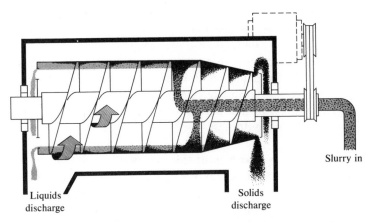

Figure 11.5 Operation of horizontal Super-D-Canter centrifuges. (Courtesy of Pennsalt Chemical Co.)

Centrifuges separate solids from liquids by sedimentation and centrifugal force. Sludge solids settle through the liquid pool and are compacted by centrifugal force against the walls of the bowl and are then conveyed by the screw conveyor to the drying or beach end of the bowl. The beach area is an inclined section of the bowl where further dewatering occurs before the solids are discharged over adjustable weirs at the opposite end of the bowl.

Bowl speed affects clarification. The clarification capacity depends upon the gravitational force and the detention time of the feed mixture in the pool. Centrifuges operate in excess of 3500 g's.

The pool volume regulates the liquid retention time and the drying beach length in the bowl. As a result, increasing the pool volume will improve solids recovery, but will decrease the cake dryness. Since the pool volume is controlled by an adjustable weir, field adjustments can be made for optimum operation. The conveyor speed can be adjusted to yield maximum cake dryness and solids recovery for a specific application.

When the feed rate to a centrifuge is increased, the retention time in the unit is decreased and the recovery decreases. Since the lower recovery results in only the removal of larger particles, a drier cake is produced. Increasing the feed solids concentration will reduce the liquid overflow from the machine, resulting in an increased recovery of solids.

Chemical flocculents (polyelectrolytes) are used to increase recovery. The flocculents both increase the structural strength of the solids and flocculate fine particles. Because of the increased removal of fine particles, chemical addition will usually lower the cake dryness.

Laboratory Experimentation While a definitive design procedure from laboratory data is not available, it is possible to determine the feasibility of centrifugation from simple laboratory tests.

The sludge to be dewatered is spun in a laboratory hand centrifuge for 30 sec at 2000 rpm (this is the maximum speed of the average hand unit). A hand centrifuge is preferred because it can be brought up to speed and stopped rapidly as compared to the automatic models. After centrifugation, the tube is tilted to determine the stability of the deposited cake. A glass stirring rod is then dropped into the cake from a distance of about $\frac{1}{2}$ cm. A fluid cake or a high degree of penetration would indicate poor recovery in continuous operation. The tests should be repeated with the addition of polymers or other conditioning chemicals. The cake solids concentration can be estimated from the volume of deposited cake in the centrifuge tube.

Performance data for the centrifugal dewatering of several sludges are summarized in Table 11.4. The relationship be-

Table 11.4. Centrifugation of Waste Sludges

Sludge	Concn., %	Cake, %	Recovery, %	Chemicals, lb/ton[a] (kg/ton[b])
Board mill	2–5	22–30	85–95	—
Kraft mill	1-5	22–34	82–95	—
Pulp	6.1	29.8	87	—
White water	3.1	42.7	99	—
Raw sludge (domestic)	—	30–40	70–90	—
Digested sludge	—	30–40	70–90	—
Primary and secondary domestic	—	15–20	85–100	10–15 (5–7.5)
Secondary domestic	—	5–15	90–100	5–10 (2.5–5)
Waste activated	1.5	5–15	90–100	5–10 (2.5–5)

[a] Short ton.
[b] Metric ton.

tween feed rate and recovery with and without the addition of polymers is shown in Fig. 11.6.

VACUUM FILTRATION

Vacuum filtration is the most common method for dewatering wastewater sludges. Vacuum filtration dewaters a slurry under applied vacuum by means of a porous media which retains the solids but allows the liquid to pass through. Media used include cloth, steel mesh, or tightly wound coil springs.

In vacuum-filter operation (Fig. 11.7), a rotary drum passes through a slurry tank in which solids are retained on the drum surface under applied vacuum. The drum submergence can vary from 12 to 60 percent. As the drum passes through the slurry, a cake is built up and water removed by filtration through the deposited solids and the filter media. The time the drum remains submerged in the slurry is the form time. As the drum emerges from the slurry tank, the deposited cake is further dried by liquid transfer to air drawn through the cake by

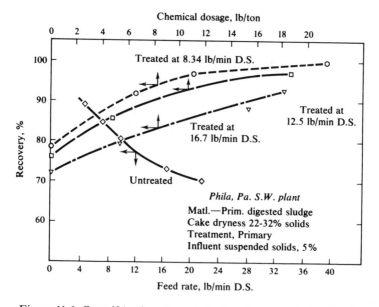

Figure 11.6 Centrifugation characteristics of sewage sludge. [*15*]

the applied vacuum. This period of the drum's cycle is called the dry time. At the end of the cycle, a knife edge scrapes the filter cake from the drum to a conveyor. The filter media is

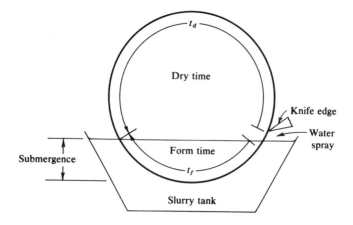

Figure 11.7 Vacuum-filter schematic.

usually washed with water sprays prior to again being im-
mersed in the slurry tank. A vacuum-filter installation is
shown in Fig. 11.8.

Figure 11.8 Vacuum-filter installation. (Courtesy of Eimco Corporation.)

The variables that influence the dewatering process are
solids concentration, sludge and filtrate viscosity, sludge compressibility, chemical composition, and the nature of the
sludge particles (size, shape, water content, etc.). [6].

The filter operating variables are vacuum, drum submergence and speed, sludge conditioning, and the type and porosity of the filter media.

The rate of filtration of sludges has been formulated according to Poiseuille's and Darcy's laws by Carmen and Coackley
[7]:

$$\frac{dV}{dt} = \frac{PA^2}{\mu\,(rcV + R_m A)} \tag{3}$$

where V = volume of filtrate

t = cycle time (approximate form time in continuous drum filters)

P = vacuum

A = filtration area

μ = filtrate viscosity

r = specific resistance

c = weight of solids/unit volume of filtrate

R_m is the initial resistance of the filter media and can usually be neglected as small compared to the resistance developed by the filter cake. The specific resistance r is a measure of the filterability of the sludge and is numerically equal to the pressure difference required to produce a unit rate of filtrate flow of unit viscosity through a unit weight of cake.

Integration and rearrangement of equation (3) yields

$$\frac{t}{V} = \frac{\mu r c}{2PA^2} V + \frac{\mu R_m}{PA} \tag{3a}$$

From equation (3a) a linear relationship will result from a plot of t/V vs. V. The specific resistance can be computed from the slope of this plot:

$$r = \frac{2bPA^2}{\mu c} \tag{4}$$

in which b is the slope of the t/V vs. V plot. The calculation of specific resistance is as shown.

Example   (see Fig. 11.9)

$$r = \frac{2PbA^2}{\mu c}$$

$$c = \frac{1}{[c_i/(100 - c_i)] - [c_f/(100 - c_f)]} = \frac{1}{95.6/4.4 - 80/20} = 0.056 \text{ g/ml}$$

$$r = \frac{(2)(526)(104.6)^2(0.004)}{(0.00895)(0.056)}$$

$$= 0.92 \times 10^6 \text{ sec}^2/\text{g}$$

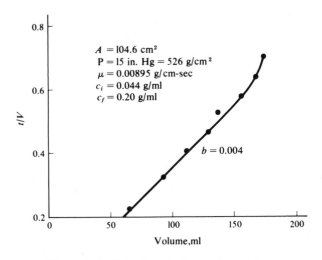

Figure 11.9 Calculation of specific resistance.

Although specific resistance has limited value for the calculation of filtration rates on drum filters, it provides a valuable tool for the evaluation of vacuum-filtration variables and

Table 11.5.

Time, sec	Vol., ml	t/V
14.5	66	0.22
29.5	92	0.32
45	112	0.40
59	129	0.46
70	134	0.52
89	156	0.57
105	167	0.63
120	174	0.69

the relative filterability of sludges. Typical values are given in Table 11.6.

Table 11.6. Specific Resistance of Some Industrial and Municipal Sludges

Sludge	Specific Resistance $\times 10^7$, sec^2/g at 500 g/cm^2	Coefficient of Compressibility, s	Reference
Neutralization of sulfuric acid with lime slurry	1–2		[8]
Neutralization of sulfuric acid with dolomitic lime slurry	3	0.77	[8]
Processing of aluminum	3	0.44	[8]
Paper industry	6	—	[8]
Neutralization of fatty acids with sodium carbonate	7	—	[8]
Froth flotation of coal	80	1.6	[8]
Malt whisky distillery	200	1.3	
Mixed chrome and vegetable tannery	300	—	[8]
Biological treatment of chemical wastes	300	—	[8]
Activated (domestic)	2880	0.81	[9]
Raw conditioned (domestic)	3.1	1.00	[9]
Digested conditioned (domestic)	10.5	—	
Digested and activated (conditioned)	14.6	1.10	[9]
Raw (domestic)	470	0.54	[9]

Most industrial-waste sludges form compressible cakes in which the filtration rate and the specific resistance is a function of the pressure difference across the cake:

$$r = r_o P^s \tag{5}$$

in which s is the coefficient of compressibility. The greater the value of s, the more compressible is the sludge. When $s = 0$, the specific resistance is independent of pressure and the sludge is incompressible.

Some generalizations on filtration characteristics can be made. Raw sewage sludge is easier to filter than digested sewage sludges, and primary sewage sludges are easier to filter than secondary sludges. Filterability is influenced by particle size, shape, and density and by the electrical charge on the particle. Smaller particles will exert a greater chemical demand than larger particles. The larger the particle size, the higher is the filter rate and the lower the cake moisture. Combinations of lime and/or ferric salts have been the most common coagulating agents used in the past. Recently, however, polyelectrolytes have proved effective coagulants in most applications. Frequently, the dual use of anionic and cationic polymers is the most economic and effective procedure. The cationic polymer effects charge neutralization and the anionic effects polymer particle bridging and agglomeration of the particles.

VACUUM-FILTRATION DESIGN

Equation (3) can be modified to express filter loading (neglecting the initial resistance of the filter media):

$$L_f = 35.7 \left(\frac{cP^{1-s}}{\mu R_0 t_f}\right)^{1/2} \tag{6}$$

where $R_0 = r \times 10^7$, sec^2/g

   P = vacuum, psi

   c = solids deposited per unit volume filtrate, g/ml

   μ = filtrate viscosity, centipoises

   t_f = form cycle time, min

Equation (6) should be modified for the prediction of filtration rates for various types of sludges:

$$L_f = 35.7 \left(\frac{P^{1-s}}{\mu R_0}\right)^{1/2} \frac{C^m}{t_f^n} \tag{7}$$

Filter operation has shown the exponent n to vary from -0.4 to -1.0. Shepman and Cornell [10] have attributed this variation to variations in cake permeability as additional cake is formed. The effect of variation in solids content fed to the filter will likewise vary m from 0.25 to 1.0. The exponents m and

n must be determined experimentally for any specific application.

Final cake moisture is related to the cake thickness, the drying time, the pressure drop across the cake, the liquid viscosity, and the air rate through the cake. The drying time will increase to a maximum beyond which very little increase will occur. In a very porous cake, the change in cake moisture with increased drying time or increased vacuum will be small, because the high air rates through the cake will cause a rapid initial drying to equilibrium. Nonporous cakes require longer drying times and high vacuum to attain a maximum cake solids content. The vacuum-filtration characteristics of several sludges are summarized in Table 11.7.

Procedure for the Design of a Vacuum Filter

Data to be Obtained
1. Volume and characteristics of sludge; variation in sludge quantity.
2. Pretreatment (thickening) and resulting concentrations of sludge to be dewatered.

Design Procedure
1. The optimum coagulant dosage is determined from a series of Büchner funnel tests and plotted as shown in Fig. 11.10a. It is important to note that the coagulant dosage must be expressed as a percentage (by weight) if the solids content of the feed to the filter is variable.
2. The design relationship is

$$L_f = 35.7 \left(\frac{P^{1-s}}{\mu R_o}\right)^{1/2} \frac{C^m}{t_f^n} \tag{7}$$

The specific resistance, R_0, is obtained from a correlation of the leaf test data. The exponents n and m are obtained from the slopes of the plots of Fig. 11.10b and d. The compressibility coefficient, s, is obtained from the slope of plot Fig. 11.10c. (*Note*: In Fig. 11.10b, C and P are held constant; in Fig. 11.10c, t_f and C are held constant; in Fig. 11.10d, t_f and P are held constant.)

Table 11.7. Vacuum Filtration Characteristics of Sludges

Sludge	Concn., %	Cake, %	Filtration Rate, lb/ft²/hr (kg/m²/hr)	Chemicals, % (by wt)	
				[a]	[b]
Primary sewage					
Undigested	6–10	66–69	6.9 (33.7)	1.0–2.0	6.0–9.0
Digested	6–10	70–73	7.2 (35.2)	2.5–3.5	7.0–12.0
High-rate trickling filter					
Mixed primary and secondary	7	68–75	7.1 (34.7)	1.5–2.5	7.0–11.0
Digested mixture	7	71	—	3.0	8.0
Activated sludge					
Mixed primary and secondary	6	75	4.5 (22.0)	2.5–3.5	5.0–10.0
Digested mixture	6	76	—	3.5	9.0
Raw primary or raw					
Primary and filter homos	—	63–72	6–20 (29–98)	0.2–1.2 [c]	
Digested primary	—	66–74	4–15 (19.5–73)	0.2–1.5	
Digested primary and activated	—	68–76	4–8 (19.5–39)	0.5–2.0	

[a] Ferric chloride.
[b] Lime.
[c] Polyelectrolyte.

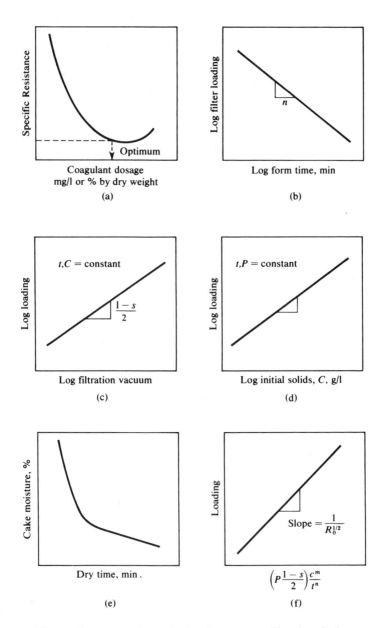

Figure 11.10 Experimental plots for vacuum-filtration design.

3. In general, the maximum initial solids concentration C is used which can be dependably expected from a prethickening unit.

4. Chemical-conditioning feed equipment is sized for the optimum dosage (on a percentage of the dry-weight solids fed to the filter).

5. It is next necessary to select a cycle time, t_c. Total cycle time on a filter may vary from 1 to 6 min. Submergence of the drum may vary from 10 to 60 percent, resulting in a maximum spread of form time from 0.1 to 3.5 min. This also yields a maximum spread in dry time of 2.5 to 4.5 min.

    a. From Fig. 11.10e a dry time is selected to give a maximum cake solids. A porous cake will crack quickly, yielding a maximum cake solids in a short dry time; other cakes will require a long dry time to achieve a maximum cake solids.

    b. A form time is next selected to give a maximum cake pickup within the cycle-time limitations established above.

6. The operating form vacuum is then selected. In general, the filter yield from a highly compressible cake will be relatively unaffected by increases in form vacuum. Filters will operate with form vacuum varying from 12 to 17 in. (30 to 43 cm) of Hg. For most economical operation, the more compressible cakes will operate at the lower form vacuum. Dry vacuum may be expected to vary from 12 to 17 in. (30 to 43 cm) of Hg.

7. Having calculated or selected the appropriate coefficients, the filter loading based on form time is computed from equation (7).

8. The form loading is converted to cycle-time loading from the relationship

$$L_c = L_f \; \frac{\% \text{ submergence}}{100} \cdot 0.8$$

The factor 0.8 compensates for that area of the filter drum where the cake is removed and the media washed.

9. The filter drum is sized:

$$\text{drum area required} = \frac{\text{lb of solids/hr fed (dry weight)}}{L \ (\text{lb/ft}^2/\text{hr})} \left(\text{or } \frac{\text{kg/hr}}{\text{kg/m}^2/\text{hr}} \right)$$

It should be noted that if the filter is to be operated less than 24 hr/day, the solids to be fed to the filter must be adjusted for the operating schedule.

PRESSURE FILTRATION

Pressure filtration, although not popular in the United States, has found widespread application in Europe. Presses have recently been automated, thereby eliminating the principal object of high labor costs. A typical filter press is shown in Fig. 11.11. Most presses are a series of plate and frame filters in which sludge is subjected to about 250 psi pressure (17.5 kg/cm^2) for a period of 30 min. to 2 hr. The filter cloth retains the solids and allows the liquid to pass. Automatic operation opens the press, discharges the cake, and washes the media between cycles.

A principal advantage of the press over a vacuum filter is the ability to achieve a drier cake (particularly when incineration is being considered).

Hydraulic presses have also been applied to further dewater filter cake from paper mill sludges for incineration. Boardmill sludge has been dewatered to 40 percent solids from 30 percent solids at a pressure of 300 psi (21 kg/cm^2) and a pressing time of 5 min.

SAND BED DRYING

For smaller sewage plants and some industrial waste-treatment plants, the most common method of sludge dewatering is drying on open or covered sand beds. Drying of the sludge occurs by percolation and evaporation. The proportion of the water removed by percolation may vary from 20 to 55 percent, depending on the initial solids content of the sludge and on the characteristics of the solids. The design and use of drying beds

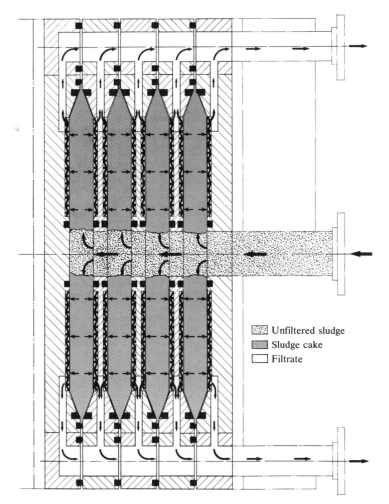

Figure 11.11 Cross-sectional view and flow diagram of plate section of a sludge filter. (Courtesy of Beloit-Passevant Co.)

are affected by climatic conditions (rainfall and evaporation).

Sludge drying beds usually consist of 4 to 9 in. (10 to 23 cm) of sand over 8 to 18 in. (20 to 45 cm) of graded gravel or stone [11]. The sand has an effective size of 0.3 to 1.2 mm and a uniformity coefficient less than 5.0. Gravel is graded from $\frac{1}{8}$ to 1 in. (0.32 to 2.54 cm). The beds are provided with underdrains spaced from 9 to 20 ft apart (2.6 to 6.1m). The under-drain piping may be vitrified clay laid with open joints having a minimum diameter of 4 in. (10 cm) and a minimum slope of about 1 percent. The filtrate is returned to the treatment plant.

Wet sludge is usually applied to the drying beds at depths of 8 to 12 in. (20 to 30 cm). Removal of the dried sludge in a "liftable state" will vary with both individual judgement and final disposal means, but usually will involve sludge of 30 to 50 percent solids. The dewatering characteristics of some sludges are shown in Fig. 11.12 and Table 11.8.

In many cases, the bed turnover can be substantially increased by the use of chemicals. Alum treatment can reduce the sludge drying time by 50 percent. The use of polymers can increase the rate of bed dewatering and also increase the depth of application. Bed yield has been reported to increase linearly with polymer dosage [12].

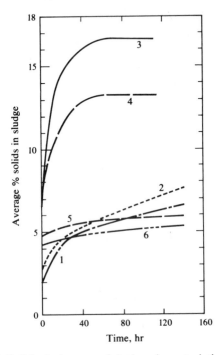

Figure 11.12 Solids drainage and drying characteristics on sand beds (numbers refer to Table 11.8)

Table 11.8. Waste-Sludge Drying Characteristics[a]

No.	Source or Type of Sludge	Solids Applied, %	Vol., %	Depth Applied, cm	Period to Liftable State, days	Solids Content Removed, %	Specific Resistance × 10^9, sec²/g (500 g/cm²)	Coefficient of Comp.
1.	Steam peeling of carrots	1.91	74.3	39.4	25	7.0	0.46	1.02
2.	Lime neutralization of pickling liquor	2.46	16.9	48.3	25	10.5	0.30	0.63
3.	Vegetable tanning	6.79	43.5	19.5	20	20	0.15	0.79
4.	Lime neutralization of regenerant liquors from ion-exchange demineralization of plating wastes	7.05	19.8	17.5	20	18	0.19	0.55
5.	Alum sludge	4.82	44.1	26.9	50	14.5	5.3	0.31
6.	Digested sewage sludge	4.28	56.8	30.5	74	25	13.5	0.60

[a] From WPRB [8].

Drying-Bed Design Drying beds have been designed on the empirical basis of square feet of bed area per capita (square meters per capita) or lb of dry solids/ft²/yr (kg/m²/yr). Values commonly employed in the United States are summarized in Table 11.9.

Table 11.9.   Sludge-Drying-Bed Design Parameters

Type of Digested Sludge	Area, ft²/capita (m²/capita)	lb of dry solids/ft²/year
Primary	1.0 (0.09)	30
Primary and standard trickling filter	1.6 (0.11)	25
Primary and activated sludge	3.0 (0.28)	20
Chemically precipitated sludge	2.0 (0.19)	22

Recently a rational method has been developed by Swanwick [*13*] based on the observed dewatering characteristics of a variety of sewage and industrial waste sludges. This procedure is outlined below:

1. Fill a 1-in. (2.54-cm)-diameter glass tube with a sand base with sludge to a depth of 12 to 18 in. (30 to 46 cm).

2. Allow complete drainage of water from the sludge (this will usually occur in a period of 12 to 48 hr, depending on the sludge moisture content and characteristic).

3. Remove the drained sludge plug and measure the moisture content after drainage.

4. Place the plug in an exposed dish for evaporation to occur. Periodically check the plug until a desired terminal moisture content is reached.

5. Measure the remaining moisture content. The difference between steps 3 and 5 is the water to be evaporated.

6. From local meteorological records, determine the annual evaporation rate and the annual rainfall rate.

7. Plot a cumulative plot of 0.75 times summation of evaporation vs. month and 0.57 times summation of rainfall vs.

month. (This is to be based on experimental evidence that the average evaporation rate from wet sludge is 75 percent that of a free water surface and that 43 percent of the rainfall is drained through the cake leaving 57 percent to be evaporated.)

8. Prepare an overlay by month, indicating the total length of time sludge discharge to a bed at that time must be held for evaporation of any rainfall plus that calculated from step 5.

9. Prepare a tabulation of the total bed area required for each month. (The required area will increase during the wet periods of the year.)

10. The design requirement will be the maximum computed in step 9.

HEAT TREATMENT

Many sludges, including sewage sludge, are difficult to dewater and require high dosages of coagulation chemicals for this purpose. Two processes have been developed which heat the sludge for short periods under pressure. This coagulates the solids, breaks down the gel structure, and reduces the hydration and hydrophilic nature of the solids. This permits rapid dewatering without the use of chemical additives.

In the Porteus process sludge is passed through a heat exchanger into a reaction vessel, where steam is injected directly into the sludge. The retention time is 30 min at 350 to 390°F (188 to 199°C) and 180 to 210 psi (12.6 to 14.7 kg/cm²). After heat treatment, the sludge passes back through the heat exchanger into a thickener-decanter. For sewage sludge, a cake from a filter press of about 40 percent moisture is obtained. The decant and press liquor is high in BOD and requires return to the treatment process.

The low-pressure Zimpro system operates with pressures in the range of 150 to 300 psi (10.5 to 21.0 kg/cm²). Sludge and air are heated by an exchanger to 250 to 300°F (149 to 167°C) before entering the reactor. Reactor temperatures are maintained at 300 to 350°F (167 to 177°C) by the injection of steam. The primary difference between Zimpro and Porteous is the injection of air to the reactor in the Zimpro process. The exhaust gases are water scrubbed. The treated sludge is disposed of by vacuum filtration or on sand beds. The filtrate normally contains 2000 to 3000 mg/l of BOD and requires biological treatment. The Zimpro process is shown in Fig. 11.13. The principal advantages of heat treatment are that the sludge is sterilized, substantially deodorized, and dewaters readily on vacuum or pressure filters.

LAND DISPOSAL

Land disposal of wet sludges can be accomplished in a number of ways: lagooning or application of liquid sludge to land by truck or spray system or by pipeline to a remote agricultural or lagoon site.

Lagooning is commonly employed for the disposal of in-organic industrial waste sludges. Sewage and organic sludges usually receive aerobic or anaerobic digestion prior to lagooning to eliminate odors and insects. Lagoons may be operated as substitutes for drying beds in which the sludge is periodically removed and the lagoon refilled. In a permanent lagoon, supernatent liquor is removed, and, when filled with solids, the lagoon is abandoned and a new site selected. Sewage sludge stored in a lagoon can be dewatered from 95 percent moisture to 55 to 60 percent moisture in a 2- to 3-yr period.

In general, lagoons should be considered where large land areas are available and the sludge will not present a nuisance to the surrounding environment.

In several cases, biological sludges after aerobic or anaerobic digestion have been sprayed on local land sites from tank wagons or pumped through agricultural pipe. Multiple applications at low dosages form a thin sludge layer which is easily worked into the soil. Reported loadings range from 100 dry tons/acre (22.4 metric tons/1000 m²) average conditions to 300 tons/acre (67.2 metric tons/1000 m²) in areas of low rainfall [14].

Excess activated sludge has been disposed of in oxidation ponds in which algal activity maintains aerobic conditions in the overlaying liquid while the sludge undergoes anaerobic digestion. This procedure has been successfully employed for municipal activated sludge at Austin, Texas, and excess activated sludge from a petrochemical plant in Houston, Texas.

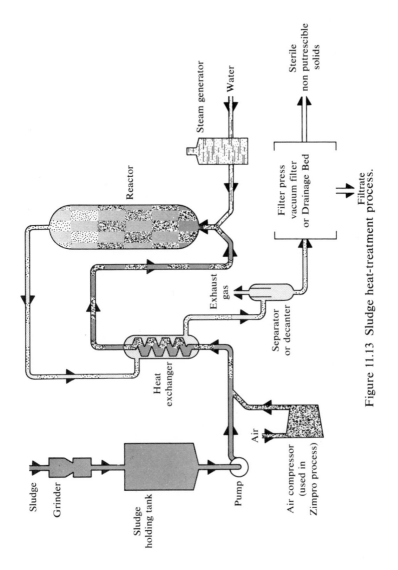

Figure 11.13 Sludge heat-treatment process.

Pipeline transportation of wet sludge for land or lagoon disposal in remote areas is gaining increasing interest, particularly for large urban communities. Sewage sludge requires
digestion prior to pumping. The relative costs of pipeline
disposal and other methods for a city of 100,000 people is
shown in Fig. 11.14.

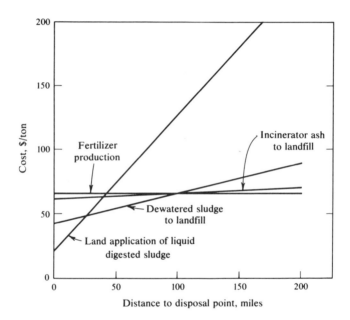

Figure 11.14 Cost of disposal of approximately 50,000 gpd of sludge
by various methods (city of 100,000 people). (After Rid-
del and Cormack[16].)

INCINERATION

After dewatering the sludge cake must be disposed of. This
can be accomplished by hauling the cake to a land-disposal
site or by incineration.

The variables to be considered in incineration are the moisture and volatile content of the sludge cake and the thermal
value of the sludge. The moisture content is of primary signi-

ficance because it will dictate whether the combustion process will be self-supporting or whether supplementary fuel will be required. The thermal value of several sludges is shown in Table 11.10.

Table 11.10. Thermal Characteristics of Waste Sludges

Sludge	Volatiles, %	Btu/lb	Kcal/Kg
Raw sewage solids	74	10,000–14,000	5550–7750
Digested sludge	70.0	5290	2940
Activated sludge	80.0	9000–10,000	5000–5550

Incineration involves drying and combustion. Various types of incineration units are available to accomplish these reactions in single or combined units. In the incineration process, the sludge temperature is raised to 212°F (100°C), at which point moisture is evaporated from the sludge. The water vapor and air temperature is increased and the temperature of the dried sludge volatiles increased to the ignition point. Some excess air is required for complete combustion of the sludge. Self-sustaining combustion is often possible with dewatered waste sludges once the burning of auxiliary fuel raises incinerator temperature to the ignition point. The primary end products of combustion are carbon dioxide, sulfur dioxide, and ash.

Incineration can be accomplished in multiple-hearth furnaces in which the sludge passes vertically through a series of hearths. In the upper hearths, vaporization of moisture occurs and cooling of exhaust gases. In the intermediate hearths, the volatile gases and solids are burned. The total fixed carbon is burned in the lower hearths. Temperatures range from 1000°F (538°C) at the top hearth to 1600 to 1800°F (870 to 980°C) at the middle hearths to 600°F (316°C) at the bottom. The exhaust gases pass through a scrubber to remove fly ash and other volatile products. A typical furnace is shown in Fig. 11.15.

In the fluidized bed, sludge particles are fed into a bed of sand fluidized by upward-moving air. A temperature of 1400 to 1500°F (760 to 815°C) is maintained in the bed, resulting

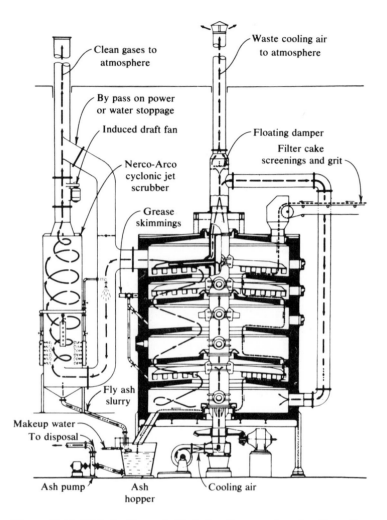

Figure 11.15 Nichols Herreshoff sludge furnace, burning flow diagram.

in rapid drying and burning of the sludge. Ash is removed from the bed by the upward-flowing combustion gases. A flow sheet for this system is shown in Fig. 11.16.

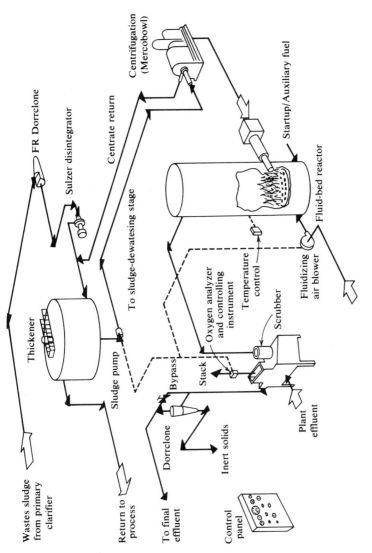

Figure 11.16 Component schematic diagram for FS disposal system. (Courtesy of Dorr Oliver, Inc.)

WET OXIDATION

Wet oxidation is a process by which the organic solids in a sludge are chemically oxidized in an aqueous phase by dissolved oxygen in a specifically designated reactor at elevated temperature and pressure [14].

The principal components of the process shown in Fig. 11.17 are a reactor, an air compressor, a heat exchanger, and a high-pressure sludge pump. The process involves pressurizing and preheating the sludge, injection of air and preheated sludge into the reactor, and oxidation in the reactor followed by gas–liquid and ash–liquid separation.

The primary design parameters are temperature, air supply, pressure, and feed solids concentration. The degree and rate of oxidation is significantly influenced by the reactor temperature. The pressure must be sufficient to condense the water vapor. Operation results have indicated COD reductions of 75 to 80 percent at temperatures of 525°F (274°C) and pressures of about 1750 psig (123 kg/cm²). The effluent liquor has a BOD between 5000 and 9000 mg/l. The residual solids can be dewatered by vacuum filtration or disposed of in lagoons or drying beds.

SLUDGE-DISPOSAL ECONOMICS

The costs of alternative methods of sewage sludge disposal as reported by Burd [1] are summarized in Table 11.11.

References

1. R. S. Burd, *A Study of Sludge Handling and Disposal*, Publ. WP-20-4, U.S. Department of Interior, FWPCA, Washington, D.C., May 1968.
2. H. J. Edde and W. W. Eckenfelder, "Theoretical Concept of Gravity Thickening; Scaling-up Laboratory Units to Prototype Design," *J. Water Pollution Control Federation*, **40**, No. 8, 1486 (1968).
3. W. J. Katz, "Sewage Sludge Thickening by Flotation," *Public Works*, **89**, No. 12, 114–115 (1958).

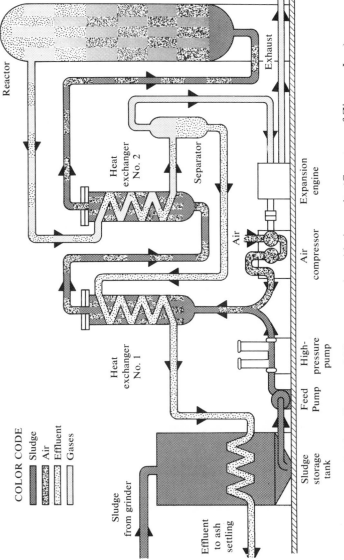

COLOR CODE

Sludge
Air
Effluent
Gases

Reactor

Exhaust

Heat exchanger No. 2

Separator

Heat exchanger No. 1

Air

Air compressor

Expansion engine

High-pressure pump

Feed Pump

Sludge storage tank

Sludge from grinder

Effluent to ash settling

Figure 11.17 Flow diagram of Zimpro wet-air-oxidation unit. (Courtesy of Zimpro, Inc.)

Table 11.11. Comparative Sludge-Handling Costs[a]

| Process | Cost, $/ton | | Operating and Capitol |
	Capital	Operating	
Thickening		2	1.50–5
Flotation		5–6	6–15[b]
Anaerobic digestion	2–2.50[c]		4–18
Vacuum filtration		8–32[b]	8–50
Centrifugation			14–28[b]
Sand-bed drying			3–20
Lagooning			2–5
Incineration	5–10	4–9	8–40
Wet oxidation		20	
Barging to sea			4–25

[a] From Burd [1].
[b] Increases with chemical addition.
[c] Per cubic foot.

4. O. F. Albertson and E. J. Guidi, "Centrifugation of Waste Sludges," paper presented before Water Pollution Control Federation, Atlantic City, N. J., October 1965.

5. R. O. Blosser, "Centrifugal Dewatering of Paper Mill Sludges," *Proc. 15th Ind. Waste Conf.*, Purdue University, Lafayette, Ind., 1960, pp. 515–528.

6. W. W. Eckenfelder, *Industrial Water Pollution Control,* McGraw-Hill, New York, 1966, pp. 236–246.

7. P. C. Coackley, in *Biological Treatment of Sewage and Industrial Wastes,* Vol II, J. McCabe and W. W. Eckenfelder, eds., Reinhold, New York, 1956.

8. Water Pollution Research, Her Majesty's Stationary Office, London, 1963.

9. W. W. Eckenfelder and D. J. O'Connor, *Biological Waste Treatment,* Pergamon Press, Oxford, 1961.

10. B. A. Shepman and C. F. Cornell, "Fundamental Operating Variables in Sewage Sludge Filtration," *Sewage and Industrial Wastes,* **28**, No. 12, 1443 (1950).

11. E. E. Seelye, *Design Data Book for Civil Engineers,* Wiley, New York, 3rd ed., 1960.

12. S. Kelman and C. P. Priesing, "Polyelectrolyte Flocculation–Sand Bed Dewatering," paper presented at Michi-

gan Conference, Water Pollution Control Federation, Detroit, June 1964.

13. Water Pollution Research, Her Majesty's Stationary Office, London, 1966.

14. *3rd Report on the Sludge of Wastewater Reclamation and Utilization,* Publ. 15, Calif. State Water Pollution Control Board, Sacramento, 1956.

15. W. J. Montanaro, "Sludge Dewatering by Centrifuges," *37th Annual Conf. WPCF,* State College, Pa., 1965.

16. M. D. Riddell and J. W. Cormack."Ultimate Disposal of Sludge in Inland Areas," *39th Annual Meeting,* Central States Water Pollution Control Assoc., 1966, Eau Claire, Wisconsin.

17. W. J. Katz and A. Geinopolos, "Sludge Thickening by Dissolved Air Flotation," *J. Water Pollution Control Federation,* **39**, 946–957 (1967).

18. W. H. Jones, "Developments with Pressurized Flotation," paper presented at 38th Annual Conference, Water Pollution Control Federation, Atlantic City, New Jersey, 1965.

12

Miscellaneous Treatment Processes

CHEMICAL COAGULATION

Inorganic and organic colloidal suspensions in wastewaters can be removed by chemical coagulation. The colloids usually found in wastewaters vary in size from approximately 100 Å to 10μ and are characterized by a zeta potential of -15 to -20 mv. The stability of the colloidal suspension is due to (1) a strong mutual repulsion induced by a high zeta potential, (2) adsorption of a relatively small lyophilic protective colloid on a larger hydrophobic colloid, or (3) adsorption of a nonionic polymer. The coagulation process[1,2], as practiced in wastewater treatment, involves destabilization of the colloidal suspension followed by flocculation to generate large particles which can subsequently be removed either by sedimentation, flotation, or filtration. For coagulation to occur, destabilization of the colloidal particles is necessary. Destabilization can be accomplished by

1. Lowering the zeta potential by the addition of a strong cationic electrolyte such as $Al_2(SO_4)_3$. This lowers the repulsive forces, permitting the van der Waals attractive forces to become effective, resulting in agglomeration. The dosage of cationic electrolyte is independent of the concentration of colloid.

2. The addition of a cationic electrolyte and an alkali resulting in the formation of charged hydrous oxides, $Me_x(OH)_y^{z+}$. The particles become adsorbed on the colloid, resulting in a coating or sheath.

3. Agglomeration by the addition of sufficient cationic polyelectrolyte to lower the zeta potential to zero. Attractive forces are then operative and mechanical bridging is achieved by the polymer. The larger the chain length, the more effective is the bridging.

4. Mutual coagulation of anionic and cationic polyelectrolytes in a system.

5. Agglomeration of a negative colloid with an anionic or nonionic polyelectrolyte.

6. Entrapment in flocs of hydrous oxide. It is believed that (2) is the principal mechanism in the coagulation of colloids in wastewater with aluminum or iron salts with some further removal by (6). To achieve optimum floc formation, the pH must be at or near the isoelectric point, which for alum is in the range 5.0 to 7.0. Sufficient alkali must be present for reaction with the aluminum (stochiometrically) and have a resulting pH at the isoelectric point. (Alkalinity in natural waters is usually in the form of HCO_3^-; when insufficient alkali is present in the wastewater it can be added as Na_2CO_3, N_aOH, or lime.)

Most effective destabilization will result from contact of the colloidal particles with small positively charged microflocs of hydrous oxide. The hydrous oxide microflocs will be generated in less than 0.1 sec, so high-intensity mixing for a short period is desirable. Following destabilization, flocculation is employed to permit the flocs to grow in size for subsequent removal from the treated wastewater. Alum and iron flocs tend to be rather fragile and easily dispersed by mixing. Activated silica at dosages of 2 to 5 mg/l is added to toughen the floc. Long-chain anionic or nonionic polymers at dosages of 0.2 to 1.0 mg/l can be added to gather and enlarge the flocs toward the end of the flocculation period. The presence of salts such as NaCl will tend to lower the zeta potential and increase the coagulant dosage. The dosage will also be increased when anionic surfactants are present in the wastewater which tend to stabilize the colloids.

Destabilization can also be accomplished by the addition of cationic polymers, which can bring the system to the isoelectric point without a change in pH. Although cationic polymers are 10 to 15 times as effective as alum as a coagulant, they are considerably more expensive. Therefore, cationic polymers serve a primary function as a trimmer after alum addition.

Engineering Considerations   From the foregoing discussion of the mechanism of coagulation, the most effective process sequence is

1. A rapid high-intensity mixing of the coagulant with the wastewater. Alkalinity, if required, can be added to the pump or piping prior to the rapid mix to ensure complete mixing prior to addition of the coagulant.

2. Activated silica and cationic polyelectrolyte, if used, should be added at the end of the rapid mix.

3. Flocculation for a period of 20 to 30 min for the development of large flocs. In this stage of the process, the objective is to obtain contact between the flocs to promote agglomeration but to maintain the mixing at a level that will not shear the flocs. Riddick [1] has recommended four flocculation bays in series with peripheral speeds tapering from 0.8 to 0.2 fps. Several writers have characterized flocculation mixing in terms of the mean velocity gradient.* Suggested values for flocculation range from 50 to 175 tapered through the basin [3]. An anionic or nonionic polyelectrolyte is added toward the end of the flocculation phase for floc gathering or agglomeration.

When coagulants are used in conjunction with flotation, the chemicals should be added downstream from the pressure-reducing valve for optimum mixing. Flocculation then occurs in the inlet chamber of the flotation unit.

Applications   Coagulation is employed for the treatment of many industrial wastes (pulp and paper, rubber, etc.) and as a polishing treatment for secondary sewage effluents (see Fig. 12.1).

The most effective coagulants and the required dosages can be determined from jar tests or by the use of zeta-potential

---

* Mean velocity gradient = $G = \sqrt{W/\mu}$, in which W = horsepower dissipated per unit volume of fluid and μ is the absolute viscosity of water. G has the units fps/ft or sec $^{-1}$.

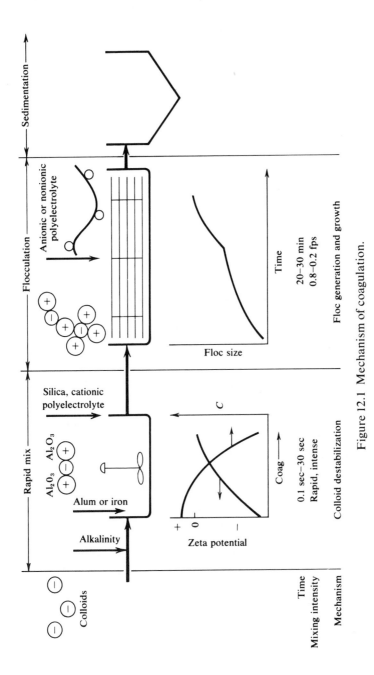

Figure 12.1 Mechanism of coagulation.

measurement. In the jar-test procedure, the optimum pH and coagulant dosage are determined by a series of tests in which pH and coagulant dosage are varied. Since optimum coagulation usually occurs at or near zero zeta potential, stepwise addition of coagulant with zeta-potential measurement will permit estimation of the dosage when the zeta potential approaches zero. These procedures have been outlined in detail [1,4].

References

1. T. M. Riddick, *Control of Colloid Stability through Zeta Potential*, Livingston Publishing Co., Wynnewood, Pa., 1968.
2. W. Stumm and C. O'Melia, "Stoichiometry of Coagulation," *J. Am. Water Works Assoc.*, **60**, No. 5, 514 (1968).
3. J. D. Walker, "High Energy Flocculation and Air and Water Backwashing," *J. Am. Water Works Assoc.*, **60**, No. 3, 32 (1968).
4. W. W. Eckenfelder, *Industrial Water Pollution Control*, McGraw-Hill, New York, 1967.

DEEP-WELL DISPOSAL

Deep-well disposal involves the injection of liquid wastes into a porous subsurface stratum which contains noncommercial brines [1]. The wastewaters are stored in sealed subsurface strata isolated from groundwater or mineral resources. Disposal wells may vary in depth from a few hundred feet (meters) to 15,000 ft (4,570 m) with capacities ranging from less than 10 to more than 2,000 gpm (38 to 7550 l/min). Wastes disposed of in wells are usually highly concentrated toxic, acidic, or radioactive or wastes high in inorganic content which are difficult or excessively expensive to treat by some other process.

The disposal system consists of the well and pretreatment equipment necessary to prepare the waste for suitable disposal into the well.

A casing, generally of steel, is cemented in place to seal the disposal stratum from other strata penetrated during drilling of the well. An injection tube transports the waste to the disposal stratum as shown in Fig. 12.2.

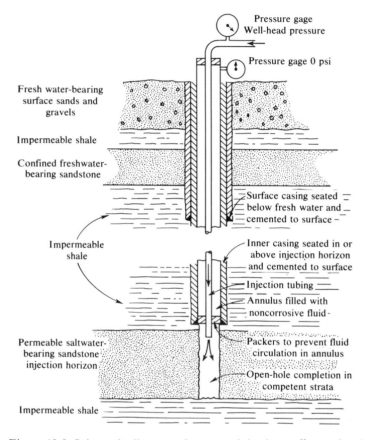

Figure 12.2 Schematic diagram of a waste-injection well completed in competent sandstone. (After Donaldson [2].)

Oil or fresh water is used to fill the annular space between the injection tube and the casing and extends to but is sealed from the injection stratum. Leaks in the injection tube or drainage to the casing can be detected by monitoring the pressure of the fluid.

The system includes a basin to level fluctuations in flow, pretreatment equipment, and high-pressure pumps. Pretreat-

ment requirements are determined by the characteristics of the
wastewater, compatibility of the wastewater and the formation
water, and the characteristics of the receiving stratum.

Pretreatment may include the removal of oils and floating
material, suspended solids, biological growths, dissolved
gases, precipitable ions, acidity, or alkalinity. A typical
system is shown in Fig. 12.3.

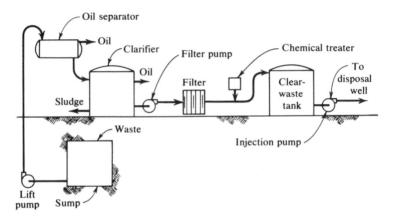

Figure 12.3 Typical subsurface waste-disposal system. (After Donaldson [2].)

The best disposal areas include sedimentary rock in the un-
fractured state, including sandstones, limestones, and dolo-
mites and unconsolidated sands. Fractured strata should be
avoided because if a vertical fissure exists groundwater con-
tamination may result.

The well-head pressure is related to the difference between
the bottom-hole pressure and the reservoir pressure. Cores of
the injection location are needed to evaluate the porosity and
permeability of the stratum and any possible reactions between
the wastewater and the stratum.

Although wastewaters should be generally free of suspended
solids, some vugular formations will accept suspended solids
without problems or an increase in injection pressure. In some
cases injection can be increased by well stimulation, which
involves the injection of mineral acids to dissolve calcium

carbonate and other acid-soluble particulates which tend to plug the stratum. Mechanical procedures involve scratching, swabbing, washing, and underreaming the well bore and shooting the uncased stratum with explosives or hydraulic fracturing.

The cost of a well system will be affected by the depth of the injection well, type of formation, geographic location, waste volume, required pretreatment, and injection pressure.

References

1. D. L. Warner, *Deep Well Injection of Liquid Waste,* Environmental Health Series, U.S. Department of Health, Education, and Welfare, Cincinnati, Ohio, April 1965.
2. E. C. Donaldson, *Subsurface Disposal of Industrial Wastes in the United States,* Bureau of Mines Inform. Circ. 8212, U.S. Department of Interior, Washington, D.C., 1964.

REDUCTION AND PRECIPITATION

Reduction and precipitation is used for the treatment of wastes containing hexavalent chromium [1, 2]. The chromium is reduced to the trivalent state by the addition of a reducing agent followed by precipitation of the reduced chromium as the hydroxide. The reactions are

$$Cr^{++++++} + \begin{matrix} \text{reducing} \\ \text{agent:} \\ Fe^{++}, \\ SO_2, \text{ or} \\ Na_2S_2O_5 \end{matrix} + H^+ \longrightarrow Cr^{+++} + \begin{matrix} Fe^{+++} \\ SO_4^{=} \end{matrix} \qquad (1)$$

$$Cr^{+++} + OH^- \longrightarrow Cr(OH)_3 \qquad (2)$$

Equation (1) is practically instantaneous at pH levels below 2.0. Although each of the reducing agents shown is effective, Fe^{++} has the disadvantages of requiring an excess dosage of about 2.5 times the theoretical and additional $Fe(OH)_3$ sludge results from neutralization. For small treatment systems,

$Na_2S_2O_5$ (sodium metabisuifite) is usually the preferred reagent. In water it hydrolyzes to $NaHSO_3$, and H_2SO_4 must be added to reduce the pH for the reducing reaction. Excess reagent must be added if dissolved oxygen is present in the wastewater.

Larger systems (batch or continuous treatment) will use SO_2, which hydrolyzes to H_2SO_3. Additional acid might not be required in many cases.

In continuous treatment systems, pH and oxidation-reduction potential control are used for acid and lime and for the SO_2 feed to the process. A typical flow diagram using automatic control is shown in Fig. 12.4.

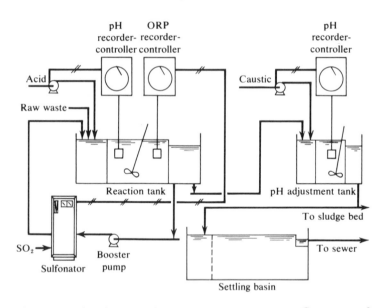

Figure 12.4 Continuous chrome-treatment system. (Courtesy of Fischer-Porter, Inc.)

Lancy [3] has developed a system where the treatment process is inserted into the plating lime.

References

1. N. S. Chamberlin and R. V. Day, *Proc. 11th Ind. Waste Conf.,* Purdue University, Lafayette, Ind. 1956.

2. W. W. Eckenfelder, *Industrial Water Pollution Control,* McGraw-Hill, New York 1967.
3. L. E. Lancy, *Sewage Ind. Wastes* **26**, No. 9, 1117 (1954).

CHLORINATION

Chlorination is employed in the treatment of wastewaters for disinfection, for the oxidation of organic compounds yielding a reduction in BOD or COD, for the oxidation of taste and odor-producing substances, and for the treatment of cyanide and other industrial wastes.

Chlorine when added to water reacts to form hypochlorous acid which subsequently dissociates:

$$Cl_2 + H_2O \longrightarrow HOCl + HCl$$
$$HOCl \rightleftharpoons H^+ + OCl^-$$

In the presence of ammonia, the HOCl reacts to form chloramines:

$$HOCl + NH_3 \rightleftharpoons NH_2Cl + H_2O \text{ (monochloramine)}$$

and

$$HOCl + NH_2Cl \rightleftharpoons NHCl_2 + H_2O \text{ (dichloramine)}$$

The proportions of monochloramine and dichloramine will be a function of the pH and the ammonia present. At low pH values, nitrogen trichloride (NCl_3) will be produced. The reactions of chlorine in water are shown in Fig. 12.5. Chlorine will initially be reduced by the presence of reducing compounds present in the wastewater. Additional chlorine results in the formation of chloramines and chloroorganic compounds. Following the oxidation of these compounds, free chlorine residual will result. This curve is frequently referred to as the breakpoint curve.

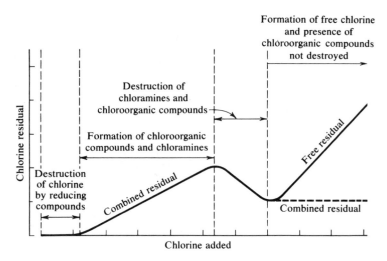

Figure 12.5 Reactions of chlorine in water.

Chlorination is universally used for the disinfection of wastewaters prior to discharge to a receiving water. Under ideal conditions, the time rate of kill follows Chicks law:

$$\frac{N}{N_0} = e^{-Kt} \tag{3}$$

in which N and N_0 are the number of viable organisms at time t and time zero, respectively, t is the time after introduction of the chlorine, and K a rate constant. K is a function of pH and temperature as well as the applied concentration of chlorine. In most cases, ideal conditions do not exist, owing to a variation in cell resistance, decrease in concentration of chlorine, and other factors, and equation (3) must be modified [1]:

$$\frac{N}{N_0} = e^{-Kt^n} \tag{4}$$

The effect of concentration of chlorine can be defined by the relationship

$$C^n t_p = \text{constant} = K \tag{5}$$

in which C is the concentration of chlorine and t_p the time required for a given percentage kill. The effectiveness of chlorination for the destruction of various organisms was shown by Berg [3] in which K varied from 0.24 for *Escherichia coli* to 6.3 for Coxsakie virus A2 for 99 percent kill at 0 to 6°C.

Chlorination is employed for the destruction of cyanide [2]. Under alkaline conditions cyanide will be oxidized in two stages:

$$CN + 2OH^- + Cl_2 \longrightarrow CNO^- + 2Cl^- + H_2O$$

and

$$CNO^- + 4OH^- + 3Cl_2 \longrightarrow 2CO_2 + N_2 + 6Cl^- + 2H_2O$$

Excess chlorine is required to complete the reaction using 7.35 parts of Cl_2/per 1 part of CN. Small waste volumes can be treated in a batch-type system. Oxidation–reduction potential can be used to control the addition of chlorine.

Chlorine has also been used for grease recovery or removal in wool-scouring wastewater. Calcium hypochlorite, $Ca(OCl)_2$, is generally used as a coagulant because this compound tends to destroy the emulsifying agent and to destabilize the colloids.

References

1. Anon., *Chlorination of Sewage and Industrial Wastes,* Manual of Practice 4, Federation of Sewage and Industrial Waste Association, 1951.
2. W. W. Eckenfelder, *Industrial Water Pollution Control,* McGraw-Hill, New York 1966, pp. 130-133.
3. G. M. Fair, J. C. Geyer, and D. A. Okun, *Water and Wastewater Engineering,* Vol. 2, Wiley, New York, 1968, 31-7, 31-12.

13

Economics of Wastewater Treatment

The costs of construction and operation of wastewater-treatment plants have recently been developed for municipal [1] and industrial [2, 3] facilities. Cost data are available for both the total plant by process (activated sludge, aerated lagoons, etc.) and for individual process units (sedimentation, neutralization, etc.). The total cost of treatment facilities can be synthesized by a summation of the costs of the individual unit processes.

All the cost data have been adjusted using either the ENR Construction Cost Index or the PHS Sewage Treatment Cost Index.

The cost of construction of municipal sewage treatment plants has been reported by Smith [1] and is summarized in Fig. 13.1. Operating cost is summarized in Fig. 13.2. The total construction costs of industrial waste treatment plants have been estimated from the following mathematical relationships:

Activated sludge:

$$C = \frac{KQ^m (S_o/S_e)^n}{k^p} \tag{1}$$

Aerated lagoons:

$$C = KQ^m (S_o/S_e)^n \tag{2}$$

292

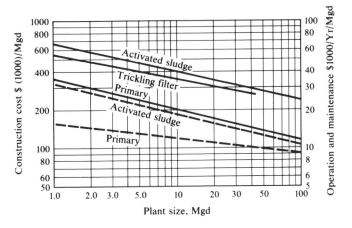

Figure 13.1 Cost relationships for a municipal sewage treatment process. (From Smith [1].)

where C = total capitol cost, thousands of dollars
Q = wastewater flow, Mgd (m³/day)
S_o = wastewater BOD, mg/l
S_e = treated effluent — BOD, mg/l
p, K, m, n = constants
k = substrate removal rate

Construction costs for unit processes treating domestic sewage are shown in Fig. 13.3.

The basis and relationships for determining the unit process cost for industrial waste treatment are summarized in Table 12.1. The cost relationship for the individual unit processes are shown in Fig. 13.4 to 13.13. In using these relationships to synthesize the total cost of a treatment plant, some additional cost figures expressed as a percentage of the total cost of construction must be added [1]: (1) contractor's profit, 10 percent; (2) contingencies, omissions, miscellaneous, piping, etc., 15 percent; (3) land acquisition, 2 percent; and (4) engineering, 6 to 12 percent. The costs of the unit processes computed from Table 13.1 should be increased by 30 to 35 percent.

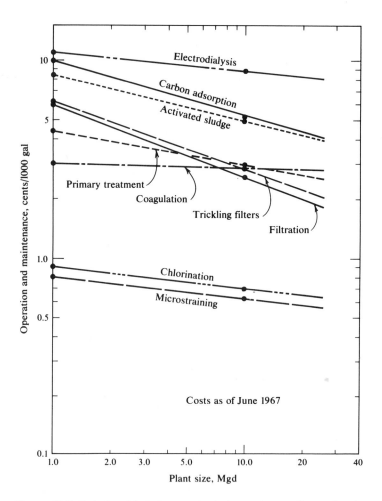

Figure 13.2 Relationship between plant size and operation and maintenance cost. (From Smith [*1*].)

The computed cost does not include the cost of land. The approximate land-area requirements for various types of waste-treatment facilities are shown in Table 13.2.

A graphical representation of the cumulative unit process cost and the associated quality of effluent for the treatment of domestic sewage is shown in Fig. 13.14 and for a chemical wastewater treatment is presented in Fig. 13.15. Similar rep-

Table 13.1.   **Summary of Bases for Unit Cost Functions**

Pretreatment or primary treatment	
Equalization	Cost vs. volume, gal
Neutralization	Cost vs. flow rate, Mgd
Oil separation	Cost/mgd vs. flow rate, Mgd
Sedimentation	Cost vs. surface area, ft²
Biological treatment	
Lagoons	Cost vs. surface area, acres
Aerated lagoons	Cost vs. volume, M gal
Activated sludge	
Aeration basin	Cost vs. volume, M gal
Final clarifier	Cost vs. surface area, ft²
Tertiary treatment	
Filtration	Cost vs. flow rate, Mgd
Ion exchange	Cost vs. flow rate, Mgd
Adsorption (carbon)	Cost vs. flow rate, Mgd
Sludge handling and disposal	
Total sludge disposal	Cost vs. flow rate, Mgd
Thickening	Cost vs. volume of thickener, gal
Flotation thickening	Cost/Mgd vs. flow rate, Mgd
Vacuum filtration	Cost vs. area of filter, ft²
Ultimate disposal	
Deep-well injection	Cost vs. flow rate, Mgd

resentations can be formulated for any process combination flow and organic concentration condition.

The following example is used to illustrate the unit-process-model and total-process-model approach in estimating the initial capital cost of a waste-treatment facility.

Table 13.2.   **Land Requirements for Wastewater Treatment**

Treatment Method	Land Requirements
Stabilization basins	3–21 acres/Mgd (0.32–2.3 hc/1000 m³/day)
Aerated lagoons	8–16 acres/Mgd (0.85–1.8 hc/1000 m³/day)
Activated sludge	1–7 acres/Mgd (0.1–0.75 hc/1000 m³/day)
Trickling filters	225–1400 ft²/Mgd (5.5–35 m²/1000 m³/day)

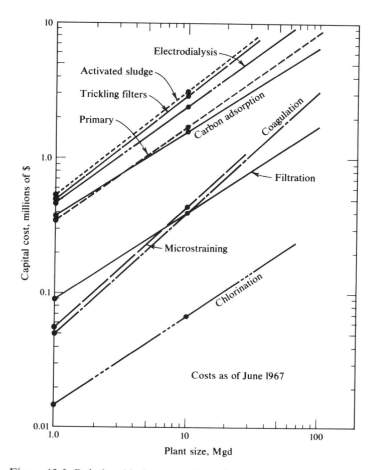

Figure 13.3 Relationship between plant size and capital cost. (From Smith [1].)

Should recycling operations be employed, an appropriate costing analysis can be formulated, balancing the savings acquired by reducing water use against the increased treatment cost in removing the contaminants attributable to recycle.

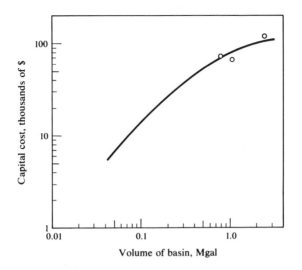

Figure 13.4 Cost vs. volume of equalization basin.

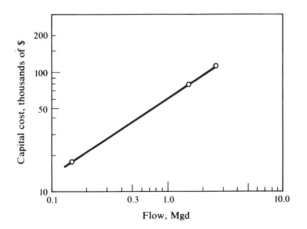

Figure 13.5 Capital-cost relationship, neutralization facilities.

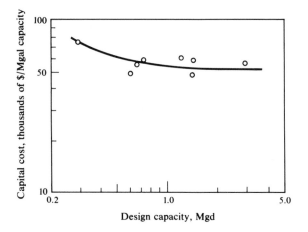

Figure 13.6 Capital-cost relationship, oil separators.

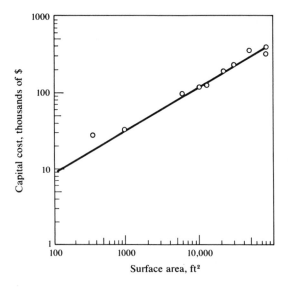

Figure 13.7 Cost vs. surface area, primary clarifier.

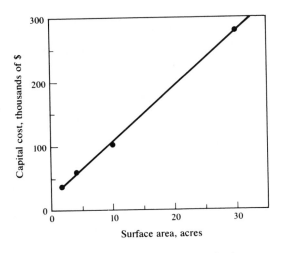

Figure 13.8 Capital-cost relationship for lagoons.

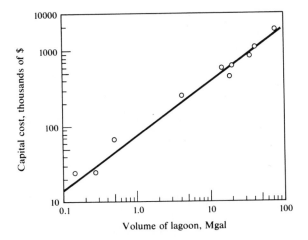

Figure 13.9 Cost vs. volume, aerated lagoon.

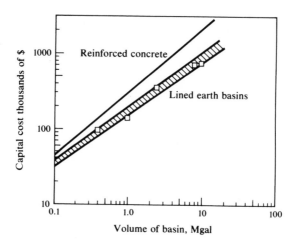

Figure 13.10 Cost vs. volume of activated sludge basin, including aerators.

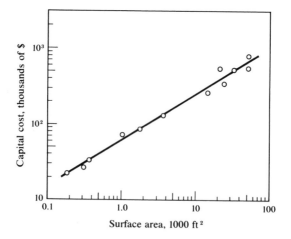

Figure 13.11 Cost vs. volume of final clarifier.

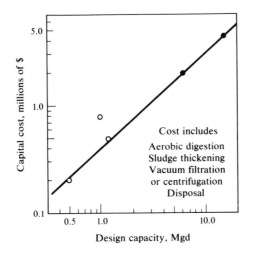

Figure 13.12 Capital-cost relationship, total sludge treatment.

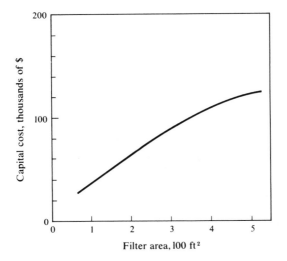

Figure 13.13 Capital-cost relationship,vacuum filters.

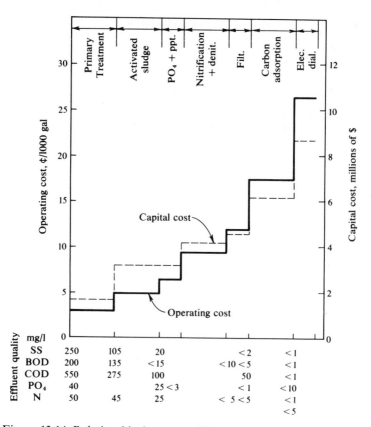

Figure 13.14 Relationship between effluent quality and capital and operating cost—muncipal sewage.

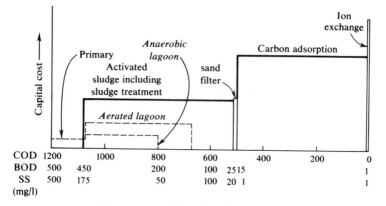

Figure 13.15 Capital cost of treatment plant vs. effluent quality vs. waste removal—chemical industrial waste.

Example: Determine the capital cost for wastewater treatment for the process alternatives. Equalization, neutralization, and sedimentation will be included as pretreatment and primary treatment.

Conditions:

$$\text{flow, } Q = 0.75 \text{ Mgd}$$
$$K = 0.002$$
$$\text{SS} = 500 \text{ mg/l}$$
$$\text{BOD} = 650 \text{ mg/l}$$

Pretreatment and primary treatment:

Equalization (1 day)		$ 65,000
Neutralization		50,000
Primary sedimentation		
Overflow rate	750 gpd/ft²	
Surface area	1000 ft²	
Suspended solids removal	85%	
BOD removal	10%	
Sludge	2650 lb/day	
Capital cost		35,000

Total cost for pretreatment and primary treatment $150,000

Activated sludge:

Aeration tank (including aeration equipment)		
of 0.322 mg (earth basin)		80,000
Final clarifiers		
Overflow rate	750 gpd/ft²	
Area	1000 ft²	
Cost		60,000

Sludge disposal:

Excess sludge:		
Primary sludge	2650 lb/day	
Secondary sludge	1395 lb/day	
Total	4045 lb/day	
Flotation thickening		34,000
Vacuum filtration 5 lb/ft²/hr, 16-hr filtration		
$\dfrac{4045}{16} \cdot \dfrac{1}{5} = 50$ ft²		25,000
Total for secondary treatment		$199,000

Total cost of facilities:

Pretreatment and primary treatment	150,000
Secondary treatment	199,000
Total	$349,000
Plus 35% for buildings, miscellaneous, piping, engineering, etc.	122,000
Total	$471,000

References

1. Robert, Smith, "Cost of Conventional and Advanced Treatment of Wastewater," *J. Water Pollution Control Federation* **40**, No. 9, 1546 (1968).
2. C. M. Rice, Eckenfelder and Associates, F. R. Weston and Resource Engineering Associates, *The Cost of Clean Water,* The Organic Chemicals Industry, report to the Federal Water Pollution Control Administration, Department of Interior, Washington, D.C. 1969.
3. James, Barnard, *The Economics of Industrial Wastewater Treatment*, M.S. Thesis, University of Texas, Austin, Texas, 1969.

Selected Readings

GENERAL

Advances in Water Pollution Research, Vol. I, II, and III, Proceedings of the International Conference held in London, September 1962, Pergamon Press, Oxford, 1964.

Advances in Water Pollution Research, Vol. I, II, and III, Proceedings of the Second International Conference held in Tokyo, August 1964, Pergamon Press, Oxford, 1965.

Advances in Water Pollution Research, Vol. I, II, and III, Proceedings of the Third International Conference held in Munich, Germany, September 1966, Water Pollution Control Federation, Washington, D. C., 1967.

Babbitt, H. E., and E. R. Baumann, *Sewerage and Sewage Treatment*, 8th ed., Wiley, New York, 1958.

Camp, T. R., *Water and Its Impurities*, Reinhold, New York, 1964.

Clark, J. W., and W. Wiessman, *Water Supply and Pollution Control*, International Textbook, Scranton, Pa., 1965.

Eckenfelder, W. W., *Industrial Water Pollution Control*, McGraw-Hill, New York, 1966.

Ehlers, V. M., and E. W. Steel, *Municipal and Rural Sanitation*, 6th ed., McGraw-Hill, New York, 1965.

Fair, G. M., J. C. Geyer, and D. A. Okun, *Water Supply and Wastewater Engineering*, Vol. I, *"Water Supply and Wastewater Removal," Wiley, New York, 1966.*

Glossary—Water and Wastewater Engineering, American Public Health Association, American Society of Civil Engineers, American Water Works Association, Water Pollution Control Federation, 1969.

McGauhey, P. H., *Engineering Management of Water Quality*, McGraw-Hill, New York, 1968.

McKinney, R. E., *Microbiology for Sanitary Engineers*, McGraw-Hill, New York, 1962.

Rich, L., *Unit Operations of Sanitary Engineering*, Wiley, New York, 1961.

Rich, L., *Unit Processes of Sanitary Engineering*, Wiley, New York, 1963.

Sawyer, C. N., and P. H. McCarty, *Chemistry for Sanitary Engineers* 2nd ed., McGraw-Hill, New York, 1967.

Standard Methods for the Examination of Water and Wastewater, 12th ed., American Public Health Association, New York, 1965.

Steele, E. W., *Water Supply and Sewerage*, 4th ed., McGraw-Hill, New York, 1960.

"Water—1968," Chemical Engineering Progress Symposium, Series 90, Vol. 64, American Institute of Chemical Engineers, New York, 1968.

GENERAL CONCEPTS OF WATER-QUALITY MANAGEMENT

Kneese, A. V., and B. T. Bower, *Managing Water Quality, Economics, Technology, Institution*, The Johns Hopkins Press, Baltimore, 1968.

Kneese, A. V., and S. C. Smith (eds.), *Water Research*, The Johns Hopkins Press, Baltimore, 1966.

McGauhey, P. H., *Engineering Management of Water Quality,* McGraw-Hill, New York, 1968.

McKee, J. E., and H. W. Wolf (eds.), *Water Quality Criteria,* 2nd ed., California Institute of Technology, Pasadena, Calif. 1963.

Proceedings of the National Symposium on Quality Standards for Natural Wastes, University of Michigan, Ann Arbor, 1966.

"Water Quality Criteria," Report of the National Technical Advisory Committee to the Secretary of the Interior, Federal Water Pollution Control Administration, Washington, D.C., 1968.

SEWAGE AND INDUSTRIAL-WASTE CHARAC-TERIZATION

Behavior of Organic Chemicals in the Aquatic Environment, A Literature Critique, Manufacturing Chemists Association, Washington, D. C., 1966.

Behavior of Organic Chemicals in the Aquatic Environment, Part II, "Behavior in Diluted Solutions," Manufacturing Chemists Association, Washington, D. C., 1968.

Biology of Water Pollution, U. S. Department of Interior, Federal Water Pollution Control Administration, Washington, D. C., 1967.

Camp, T. R., *Water and Its Impurities*, Reinhold, New York, 1964.

Department of Scientific and Industrial Research, Water Pollution Research (annual report), Her Majesty's Stationery Office, London.

Faust, S. D., and J. V. Hunter (eds.), *Principles and Applications of Water Chemistry*, Wiley, New York, 1967.

Klein, L., *River Pollution: 1. Chemical Analysis*. Butterworth Inc., Washington, D.C. 1959.

Sawyer, C. N., and P. H. McCarty, *Chemistry for Sanitary Engineers*, 2nd ed., McGraw-Hill, New York, 1967.

Standard Methods for the Examination of Water and Wastewater, 12th ed., American Public Health Association, New York, 1965.

ANALYSIS OF POLLUTIONAL EFFECTS IN NATU-RAL WATERS

Advances in Water Pollution Research, Vol. I and III, Proceedings of the International Conference held in London, September 1962, Pergamon Press, Oxford, 1964.

Advances in Water Pollution Research, Vol. I and III, Proceedings of the Third International Conference held in Tokyo, August 1964, Pergamon Press, Oxford, 1965.

Advances in Water Pollution Research, Vol. I and III, Proceedings of the Third International Conference held in Munich, Germany, September 1966, Water Pollution Control Federation, Washington, D.C., 1967.

Department of Scientific and Industrial Research, Water
   Pollution Research (annual report), Her Majesty's Sta-
   tionery Office, London.
Gloyna, E. F., and W. W. Eckenfelder (eds.), *Advances in
   Water Quality Improvement*, Vol. I, University of Texas
   Press, Austin, 1968.
Heukelekian, H., and N. C. Dondero (eds.), *Principles and
   Applications in Aquatic Microbiology*, Wiley, New York,
   1964.
Klein, L., with J. R. E. Jones, H. A. Hawkes, and A. L.
   Downing, *River Pollution: 2. Causes and Effects*, But-
   terworth Inc., Washington, D.C. 1962.
Klein, L., *River Pollution: 3. Control*, Butterworth Inc., Wash-
   ington, D.C., 1966.
Oxygen Relationships in Streams, R. A. Taft Center Tech-
   nical Report, W58-2, U.S. Department of Health, Edu-
   cation, and Welfare, Washington, D.C., 1958.

CHARACTERISTICS OF MUNICIPAL SEWAGE

Babbitt, H. E., and E. R. Baumann, *Sewerage and Sewage
   Treatment*, 8th ed., Wiley, New York, 1958.
Department of Scientific and Industrial Research, Water
   Pollution Research (annual report), Her Majesty's Sta-
   tionery Office, London.
Steele, E. W., *Water Supply and Sewerage*, 4th ed., Mc-
   Graw-Hill, New York, 1960.

INDUSTRIAL WASTES

Besselievre, E. B., *The Treatment of Industrial Wastes,*
   McGraw-Hill, New York, 1969.
Department of Scientific and Industrial Research, Water
   Pollution Research (annual report), Her Majesty's Sta-
   tionery Office, London.
Eckenfelder, W. W., *Industrial Water Pollution Control*,
   McGraw-Hill, New York, 1966.
Gloyna, E. F., and W. W. Eckenfelder, (eds.), *Advances in
   Water Quality Improvement*, Vol. I, University of Texas
   Press, Austin, 1968.

Gurnham, C. F., *Principles of Industrial Waste Treatment,* Wiley, New York, 1955.

Industrial Wastewater Control, Vol. II, "Chemical Technology," C. F. Gurnham ed., Academic Press, New York, 1965.

Isaac, P. C. G. (ed.), *Waste Treatment,* Pergamon Press, Oxford, 1960.

Nemerow, N., *Theories and Practices of Industrial Waste Treatment,* Addison-Wesley, Reading, Mass., 1963.

Proceedings of the Purdue Industrial Waste Conferences, Engineering Extension Series, Purdue University, Lafayette, Ind.

Reuse of Water in Industry, Butterworth, London, 1963.

Ross, R. D. (ed.), *Industrial Waste Disposal,* Reinhold, New York, 1968.

The Cost of Clean Water, Vol. III, "Industrial Waste Profiles;" No. 1, Blast Furnaces and Steel Mills; No. 2 Motor Vehicles and Parts; No. 3, Paper Mills; No. 4, Textile Mill Products; No. 5, Petroleum Refining; No. 6, Canned and Frozen Fruits and Vegetables; No. 7, Leather Tanning and Finishing; No. 8, Meat Products; No. 9, Dairies; No. 10, Plastic Materials and Resins; U. S. Department of Interior, Federal Water Pollution Control Administration, Washington, D.C., 1968.

"Water—1968." Chemical Engineering Progress Symposium Series 90, Vol. 64, American Institute of Chemical Engineers, New York, 1968.

"Water Reuse," Chemical Engineering Progress Symposium Series 78, Vol. 63, American Institute of Chemical Engineers, New York, 1967.

"Water Technology in the Pulp and Paper Industry," TAPPI Monograph Series No. 18, Technical Association of the Pulp and Paper Industry, New York, 1957.

WASTEWATER-TREATMENT PROCESSES

Babbitt, H. E., and E. R. Baumann, *Sewerage and Sewage Treatment,* 8th ed., Wiley, New York, 1958.

Camp, T. R., *Water and Its Impurities,* Reinhold, New York, 1964.

Clark, J. W., and W. Wiessman, *Water Supply and Pollution Control*, International Textbook, Scranton, Pa., 1965.

Eckenfelder, W. W., *Industrial Water Pollution Control*, McGraw-Hill, New York, 1966.

Eckenfelder, W. W., and D. L. Ford, "Laboratory and Design Procedures for Wastewater Treatment Processes," *Technical Report CRWR-31*, Center for Research in Water Resources, The University of Texas, Austin, 1969.

Fair, G. M., J. C. Geyer, and D. A. Okun, *Water and Wastewater Engineering*, Vol. I "Water Supply and Wastewater Removal," Wiley, New York, 1966.

Gloyna, E. F. and W. W. Eckenfelder *Advances in Water Quality Improvement,* Vol. I, University of Texas Press, Austin, 1968.

Isaac, P. C. G. (ed.), *Waste Treatment*, Pergamon Press, Oxford, 1960.

PRETREATMENT AND PRIMARY TREATMENT

Advances in Water Pollution Research, Vol. I, II, and III, Proceedings of the International Conference held in London, September 1962, Pergamon Press, Oxford, 1964.

Advances in Water Pollution Research, Vol. I, II, and III, Proceedings of the Second International Conference held in Tokyo, August 1964, Pergamon Press, Oxford, 1965.

Advances in Water Pollution Research, Vol. I, II, and III, Proceedings of the Third International Conference held in Munich, Germany, September 1966, Water Pollution Control Federation, Washington, D.C., 1967.

Babbitt, H. E., and E. R. Baumann, *Sewerage and Sewage Treatment,* 8th ed., Wiley, New York, 1958.

Clark, J. W. and W. Wiessman, *Water Supply and Pollution Control*, International Textbook, Scranton, Pa., 1965.

Department of Scientific and Industrial Research, Water Pollution Research (annual report), Her Majesty's Stationery Office, London.

Eckenfelder, W. W., *Industrial Water Pollution Control*, McGraw-Hill, New York, 1966.

Fair, G. M., J. C., Geyer, and D. A. Okun, *Water and Wastewater Engineering,* Vol. I. "Water Supply and Wastewater Removal," Wiley, New York, 1966.

Isaac, P. C. G. (ed.), *Waste Treatment*, Pergamon Press, Oxford, 1960.

Proceedings of the Purdue Industrial Waste Conferences, Engineering Extension Series, Purdue University, Lafayette, Ind.

Rich, L., *Unit Processes of Sanitary Engineering*, Wiley, New York, 1963.

"Sewage Treatment Plant Design," *Manual of Practice No. 8,* Water Pollution Control Federation, Washington, D.C., 1959.

OXYGEN TRANSFER AND AERATION

Department of Scientific and Industrial Research, Water Pollution Research (annual report), Her Majesty's Stationery Office, London.

Eckenfelder, W. W., *Industrial Water Pollution Control*, McGraw-Hill, New York, 1966.

Eckenfelder, W. W., and D. L. Ford, "Laboratory and Design Procedures for Wastewater Treatment Processes." *Technical Report CRWR-31*, Center for Research in Water Resources, The University of Texas, Austin, 1969.

Eckenfelder, W. W., and D. J. O'Connor, *Biological Waste Treatment*, Macmillan, New York, 1961.

Gloyna, E. F., and W. W. Eckenfelder, *Advances in Water Quality Improvement*, Vol. I, University of Texas Press, Austin, 1968.

BIOLOGICAL TREATMENT

Advances in Water Pollution Research, Vol. II, Proceedings of the International Conference held in London, September 1962, Pergamon Press, Oxford, 1964.

Advances in Water Pollution Research, Vol. II, Proceedings of the Second International Conference held in Tokyo, August 1964, Pergamon Press, Oxford, 1965.

Advances in Water Pollution Research, Vol. II, Proceedings of the Third International Conference held in Munich, Germany, September 1966, Water Pollution Control Federation, Washington, D.C., 1967.

Chow, V. T. (ed.), *Advances in Hydroscience*, Vol. III. pp. 154–189, Academic Press, New York, 1966.

Eckenfelder, W. W., *Industrial Water Pollution Control*, McGraw-Hill, New York, 1966.

Eckenfelder, W. W., and D. L. Ford, "Laboratory and Design Procedures for Wastewater Treatment Processes," *Technical Report CRWR-31*, Center for Research in Water Resources, The University of Texas, Austin, 1969.

Eckenfelder, W. W., and B. J. McCabe, *Advances in Biological Waste Treatment*, Macmillan, New York, 1963.

Eckenfelder, W. W., and D. J. O'Connor, *Biological Waste Treatment*, Macmillan, New York, 1961.

Fair, G. M., J. C. Geyer, and D. A. Okun, *Water and Wastewater Engineering,* Vol. I, "Water Supply and Wastewater Removal," Wiley, New York, 1966.

Gloyna, E. F., and W. W. Eckenfelder, *Advances in Water Quality Improvement,* Vol. I, University of Texas Press, Austin, 1968.

Isaac, P. C. G. (ed.). *Waste Treatment*, Pergamon Press, Oxford, 1960.

McCabe, B. J., and W. W. Eckenfelder, *Biological Treatment of Sewage and Industrial Wastes*, Vol. II, *Anaerobic Digestion and Solids-Liquid Separation."* Reinhold, New York, 1958.

Rich, L., *Unit Processes of Sanitary Engineering,* Wiley, New York, 1963.

"Sewage Treatment Plant Design," *Manual of Practice No. 8*, Water Pollution Control Federation, Washington, D.C., 1959.

The Activated Sludge Process in Sewage Treatment—Theory and Application, University of Michigan Press, Ann Arbor, 1966.

TERTIARY TREATMENT

Advances in Water Pollution Research, Vol. II, Proceedings of the International Conference held in London, September 1962, Pergamon Press, Oxford, 1964.

Advances in Water Pollution Research, Vol. II, Proceedings of the Second International Conference held in Tokyo, August 1964, Pergamon Press, Oxford, 1965.

Advances in Water Pollution Research, Vol. II, Proceedings of the Third International Conference held in Munich, Germany, September 1966, Water Pollution Control Federation, Washington, D.C., 1967.

Summary Report—Advanced Waste Treatment Research, U.S. Public Health Service Publication No. 999-WP-24, April 1965.

"Water—1968," Chemical Engineering Progress Symposium Series 90, Vol. 64, American Institute of Chemical Engineers, New York, 1968.

Water Pollution Control Research Series, "Advanced Waste Treatment Research," *AWTR 1–18*, U. S. Department of Interior, Federal Water Pollution Control Administration, Washington, D.C., 1964–1967.

SLUDGE HANDLING AND DISPOSAL

Advances in Water Pollution Research, Vol. II, Proceedings of the International Conference held in London, September 1962, Pergamon Press, Oxford, 1964.

Advances in Water Pollution Research, Vol. II, Proceedings of the Second International Conference held in Tokyo, August, 1964, Pergamon Press, Oxford, 1965.

Advances in Water Pollution Research, Vol. II, Proceedings of the Third International Conference held in Munich, Germany, September 1966, Water Pollution Control Federation, Washington, D.C., 1967.

Babbitt, H. E., and E. R. Baumann, *Sewerage and Sewage Treatment*, 8th ed., Wiley, New York, 1958.

Burd, R. S., "A Study of Sludge Handling and Disposal, "*Publication WP-20-4*, U.S. Department of Interior,

Federal Water Pollution Control Administration, Washington, D.C., 1968.

Fair, G. M., J. C. Geyer, and D. A. Okun, *Water and Wastewater Engineering,* Vol. I, "Water Supply and Wastewater Removal," Wiley, New York, 1966.

McCabe, B. J., and W. W. Eckenfelder, *Biological Treatment of Sewage and Industrial Wastes*, Vol. II, *"Anaerobic Digestion and Solids-Liquid Separation."* Reinhold, New York, 1958.

"Sewage Treatment Plant Design," *Manual of Practice No. 8*, Water Pollution Control Federation, Washington, D.C., 1959.

MISCELLANEOUS TREATMENT PROCESSES

Department of Scientific and Industrial Research, Water Pollution Research (annual report), Her Majesty's Stationery Office, London.

Eckenfelder, W. W. *Industrial Water Pollution Control*, McGraw-Hill, New York, 1966.

Fair, G. M., J. C. Geyer, and D. A. Okun, *Water and Wastewater Engineering,* Vol. I, "Water Supply and Wastewater Removal," Wiley, New York, 1966.

Nemerow, N., *Theories and Practices of Industrial Waste Treatment* Addison-Wesley, Reading, Mass., 1963.

Rich, L., *Unit Processes of Sanitary Engineering*, Wiley, New York, 1963.

Summary Report—Advanced Waste Treatment Research, U.S. Public Health Service Publication No. 999-WP-24, April 1965.

"Water—1968," Chemical Engineering Progress Symposium, Series 90, Vol. 64, American Institute of Chemical Engineers, New York, 1968.

ECONOMICS OF WASTEWATER TREATMENT

Davis, R. K., *The Range of Choice in Water Management,*
The Johns Hopkins Press, Baltimore, 1968.

The Cost of Clean Water, Vol. III, "Industrial Waste
Profiles": No. 1, Blast Furnaces and Steel Mills; No. 2,
Motor Vehicles and Parts; No. 3, Paper Mills; No. 4,
Textile Mill Products; No. 5, Petroleum Refining; No. 6,
Canned and Frozen Fruits and Vegetables; No. 7,
Leather Tanning and Finishing; No. 8, Meat Products;
No. 9, Dairies; No. 10, Plastic Materials and Resins; U.S.
Department of Interior, Federal Water Pollution Control
Administration, Washington, D.C., 1968.

Index

COMMON EQUIVALENTS

Approximate Common Equivalents

1 inch	= 25 millimeters
1 foot	= 0.3 meter
1 yard	= 0.9 meter
1 mile	= 1.6 kilometers
1 square inch	= 6.5 sq centimeters
1 square foot	= 0.09 square meter
1 square yard	= 0.8 square meter
1 acre	= 0.4 hectare†
1 cubic inch	= 16 cu centimeters
1 cubic foot	= 0.03 cubic meter
1 cubic yard	= 0.8 cubic meter
1 quart (lq)	= 1 liter†
1 gallon	= 0.004 cubic meter
1 ounce (avdp)	= 28 grams
1 pound (avdp)	= 0.45 kilogram
1 horsepower	= 0.75 kilowatt
1 millimeter	= 0.04 inch
1 meter	= 3.3 feet
1 meter	= 1.1 yards
1 kilometer	= 0.6 mile
1 sq centimeter	= 0.16 square inch
1 square meter	= 11 square feet
1 square meter	= 1.2 square yards
1 hectare†	= 2.5 acres
1 cu centimeter	= 0.06 cubic inch
1 cubic meter	= 35 cubic feet
1 cubic meter	= 1.3 cubic yards
1 liter†	= 1 quart (lq)
1 cubic meter	= 250 gallons
1 gram	= 0.035 ounces (avdp)
1 kilogram	= 2.2 pounds (avdp)
1 kilowatt	= 1.3 horsepower

†common term not used in SI